The Ethics of Time

Bloomsbury Studies in Continental Philosophy

This series presents cutting edge scholarship in the field of modern European thought. The wholly original arguments, perspectives and research findings in titles in this series make it an important and stimulating resource for students and academics from across the discipline.

Breathing with Luce Irigaray, edited by Lenart Skof and Emily A. Holmes
Deleuze and Art, Anne Sauvagnargues
Deleuze and the Diagram: Aesthetic Threads in Visual Organization, Jakub Zdebik
Derrida, Badiou and the Formal Imperative, Christopher Norris
Desire in Ashes: Deconstruction, Psychoanalysis, Philosophy, edited by Simon Morgan Wortham and Chiara Alfano
Early Phenomenology, edited by Brian Harding and Michael R. Kelly
Egalitarian Moments, Devin Zane Shaw
Ernst Bloch and His Contemporaries, Ivan Boldyrev
Why there is no Post-Structuralism in France, Johannes Angermuller
Gadamer's Poetics: A Critique of Modern Aesthetics, John Arthos
Heidegger, History and the Holocaust, Mahon O'Brien
Heidegger and the Emergence of the Question of Being, Jesús Adrián Escudero
Husserl's Ethics and Practical Intentionality, Susi Ferrarello
Immanent Transcendence: Reconfiguring Materialism in Continental Philosophy, Patrice Haynes
Merleau-Ponty's Existential Phenomenology and the Realization of Philosophy, Bryan A. Smyth
Mortal Thought: Hölderlin and Philosophy, James Luchte
Nietzsche and Political Thought, edited by Keith Ansell-Pearson
Nietzsche as a Scholar of Antiquity, Helmut Heit
Philosophy, Sophistry, Antiphilosophy: Badiou's Dispute with Lyotard, Matthew R. McLennan
The Poetic Imagination in Heidegger and Schelling, Christopher Yates
Post-Rationalism: Psychoanalysis, Epistemology, and Marxism in Post-War France, Tom Eyers
Revisiting Normativity with Deleuze, edited by Rosi Braidotti and Patricia Pisters
Towards the Critique of Violence: Walter Benjamin and Giorgio Agamben, Brendan Moran and Carlo Salzani

This engraving comes from the *Regula emblematica Sancti Benedicti*, dated in 1783. It suggests that in the heart of things there is an entrapment—the original meaning of "scandal" is that of a mechanism of entrapment. In this book, I make the claim that time, alluded here by various parts of a clock lying next to the open heart, *is* this scandal. The inscription over the engraving reads *INTIMA PANDIT*, which in one translation can be rendered as "the innermost opens up." In due time, time—this "innermost intimacy"—reveals itself. Ethics is perhaps nothing else than this self-revelation of time in time. JPM

The Ethics of Time

A Phenomenology and Hermeneutics of Change

John Panteleimon Manoussakis

BLOOMSBURY ACADEMIC
LONDON • NEW YORK • OXFORD • NEW DELHI • SYDNEY

BLOOMSBURY ACADEMIC
Bloomsbury Publishing Plc
50 Bedford Square, London, WC1B 3DP, UK
1385 Broadway, New York, NY 10018, USA

BLOOMSBURY, BLOOMSBURY ACADEMIC and the Diana logo are
trademarks of Bloomsbury Publishing Plc

First published in Great Britain 2017
Paperback edition first published 2018

A catalogue record for this book is available from the British Library.

ISBN: HB: 9781474299169
PB: 9781350101814
ePDF: 9781474299152
ePub: 9781474299176

A catalog record for this book is available from the Library of Congress.

Series: Bloomsbury Studies in Continental Philosophy

Typeset by RefineCatch Limited, Bungay, Suffolk.

For Richard Kearney and Jean-Luc Marion,
my teachers and mentors who showed me the way.

Zeus, who leads mortals to understanding,
Ordering that wisdom arises from suffering,
Drips over a slumbering mind
Painful memories,
So that, even against our will,
We may become wise.
It is a violent grace that gods sent foth
Seated upon their awful thrones.

Aeschylus, *Agamemnon*, 176–183

Contents

List of Abbreviations

Works by Gregory of Nyssa

An et res	Gregory of Nyssa, De Anima et resurrectione, GNO III
Cant	Gregory of Nyssa, In canticum canticorum, GNO VI
Eun	Gregory of Nyssa, Contra Eunomium libri III, GNO II
Inscr	Gregory of Nyssa, In inscriptiones Psalmorum, GNO V
Op Hom	Gregory of Nyssa, De hominis opificio, PG 44
Or Dom	Gregory of Nyssa, De oratione dominica, PG 44

Works by Maximus the Confessor

Amb	Maximus the Confessor, Ambigua ad Thomam, PG 91
Car	Maximus the Confessor, Capita de Caritate, PG 90; Philokalia II
DP	Maximus the Confessor, Disupatio cum Pyrrho, PG 91
Epp	Maximus the Confessor, Epistulae, PG 91
Opusc	Maximus the Confessor, Opuscula theologica et polemica, PG 91
Or Dom	Maximus the Confessor, Expositio orationis dominicae, Philokalia II
The Oec	Maximus the Confessor, Capita theologica et oeconomica, PG 90; Philokalia II

Works by Søren Kierkegaard

CA	S. Kierkegaard, *The Concept of Anxiety*. Edited and translated by Reidar Thomte. Princeton: Princeton University Press, 1980.
EO	S. Kierkegaard, *Either/Or*. Edited and translated by Howard V. Hong and Edna H. Hong. Princeton: Princeton University Press, 1987.
JP	*Søren Kierkegaard's Journals and Papers*. Edited and translated by Howard V. Hong and Edna H. Hong. Bloomington: Indiana University Press, 1967–1978.

PF S. Kierkegaard, *Philosophical Fragments*. Edited and translated by Howard V. Hong and Edna H. Hong. Princeton, Princeton University Press, 1985.

SUD S. Kierkegaard, *The Sickness Unto Death*. Edited and translated by Howard V. Hong and Edna H. Hong, Princeton: Princeton University Press, 1980.

Editions

ANF *Ante-Nicene Fathers*, edited by Alexander Roberts. Peabody: Hendrickson Publishers, 2004.

ΒΕΠΕΣ Βιβλιοθήκη Ἑλλήνων Πατέρων καὶ Ἐκκλησιαστικῶν Συγγραφέων. Athens: Ἀποστολική Διακονία, 1955–.

GNO Werner W. Jaeger et al., *Gregorii Nysseni Opera* (Leiden: Brill, 1960–1996).

NPNF *Nicene and Post-Nicene Fathers*, edited by Philip Schaff and Henry Wace. Buffalo, NY: Christian Literature Publishing, 1893.

PG J. P. Migne, *Patrologia Cursus Completus, Series Graeca* (Paris: 1857–1866).

Philokalia Φιλοκαλία τῶν Ἱερῶν Νηπτικῶν, I–V. Athens: Aster, 1984.

Preface

This book has had a long period of gestation. But then again, what is the right time to write about time? Is time ever on time?

The subject of this book is time—but time understood as movement and change: what the ancients knew as the problem of *kinesis*. Movement finds here a twofold application. First, as a cosmology of movement: considering the world's becoming, that is, history, through the notion of motion and, secondly, as an anthropology of movement. These two distinct levels, the macrocosmic and the microcosmic, are not unrelated. Movement is somehow—and a great deal of this book sheds lights on this "somehow"—an agent of ethics, broadly understood, as having something to do with the notion of the good. One could speak of a cosmological conversion as much as of a conversion of one's heart and mind. Narratives of conversion, and among them the most well-known, that of St. Augustine's *Confessions*, seem to relate these two levels with each other and with the ways in which the good appears "as the result of" time—if saying this does not, indeed, suggest too much too soon.

To take a familiar example: Ebenezer Scrooge's character transformation in the course of a single night cannot be explained apart from the visitation of the ghosts of Christmas past, present, and future. I take Dickens's classic story as an apt illustration of the ethical implications that the opening of the temporal horizon might have upon a character. The moral of Dickens's story is not without its precedents in the literary canon; Oedipus's encounter with the Sphinx at the gates of Thebes represents a similar example of time breaking into the present by means of the outworldly and the uncanny. What the ghosts of the Christmas past, present, and past were for Scrooge is the Sphinx's enigma for Oedipus: namely, an invitation to consider oneself through the perspective of time. As we shall see in chapter 5, Oedipus, like Dickens's character, was living in a present immune to the aftermaths of the after. In both cases, it is not so much the case that the eternal breaks into time, as in Kierkegaard's understanding of the moment, but rather the case of time coming to interrupt the illusion of a fake eternity sustained by the now. So, when Don Giovanni, in the final act from Mozart's opera, gives his hand to the ghost of the Commendatore, thereby sealing his fate (*Dammi la mano in pegno!*), are we not to recognize in this gesture the extension of a past that reaches out to the future, inasmuch as the Commendatore's grasping of Don Giovanni's hand was already prefigured in Don Giovanni's seductive caressing of Zerlina's hand (*Là ci darem la mano*)? Does not the entire opera unfold between these two hand gestures, between the seductive caress and the condemning grasp? Does it not take place in the place opened up between a before and an after? And is this temporal horizon not precisely the place of ethics?

* * *

This book is comprised of two large parts: the first, *Prolegomena to Any Future*, spells out in a quite abstract way the theoretical presuppositions of movement with respect to the two interrelated levels already mentioned: the cosmological (chapter 1) and the anthropological (chapter 2). Chapter 1 traces the notion of *kinesis* through three thinkers, each responsible for a ground-breaking advancement in the philosophy of movement, namely, Anaxagoras, Maximus the Confessor, and Kierkegaard. This chapter maps in detail the philosophical genealogy of time's vindication against the negative Platonic and Neoplatonic views, according to which time and history are seen as the byproducts of being's fall into materiality. Chapter 2 in a sense repeats the task of chapter 1 by discovering time within the life of consciousness. This second chapter follows the work of Bergson, Husserl, Heidegger (the analysis of boredom), and Levinas (the analysis of indolence). It proceeds, then, to discuss what I call the eschatological constitution of consciousness in the work of Marion, and it concludes with raising the question of the body's connection to time, as provoked by a phenomenological reading of Augustine and Freud. The analysis of the *Prolegomena* anticipates a great deal of the discussion in the chapters to follow.

The second part of the book, *The Scandal of the Good*, follows a quasi-narrative pattern, where the theoretical work laid down by the *Prolegomena* is applied more concretely to history or, at least, to a historical development structured around three gardens and hence three respective sections: *Being at the Beginning* (chapters 3 to 5), *Being Divided* (chapters 6 to 8), and *Final Being* (chapters 9 and 10).

In a certain sense, and if one were to consider these three gardens in the absence of time, they would have appeared as only one. Yet the fact that, *sub species temporis*, this one garden becomes three is of cardinal importance to my thesis. The three gardens are borrowed from St. Augustine's *Confessions* (in his narrative they are located in Thagaste, Milan, and Ostia) and they, in turn, mirror three viewpoints in the salvific history: the Garden of Eden (creation), the Garden of Gethsemane (redemption), and the eschatological Garden of Paradise (salvation)—or, simply, past, present, and future. In this ambitious journey from the beginning to the end, Augustine serves as my Virgil, reading the *Confessions* closely, but without the intention of providing a commentary on them. There are also literary and cinematic excursions: a reading of Sophocles' *Oedipus Rex* in chapter 5, a discussion of Lars von Trier's *Dogville* in chapter 8, and, for the finale in chapter 10, the conjoining of Pirandello's creativity to that of the Taviani brothers.

Put abstractly (and it would have been very tedious to write a book such as this without encasing it in some familiar narrative), the argument made through these eight chapters is this: there is no change except in time—a time made possible by the givenness of a consciousness that is eschatologically orientated. Whether on the cosmological or the anthropological level, the good is what at the moment (that is, viewed only through the perspective of the moment) might appear as its opposite: imperfection and evil. Or, put otherwise, evil is a moment in the temporalized unfolding of the good. Thus, I have resisted the Manichean temptation of making the good the opposite of evil, especially as if good and evil were two contemporaneous poles in an opposition of synchronicity.

In many ways, this book is a sequel to my *God After Metaphysics*, in which I tried to formulate a "theological aesthetics," as in this one—the second volume of a trilogy to be—my aim is the articulation of a "phenomenological ethics."

I understand that speaking of good and evil, I position myself in conversation with past attempts to formulate an answer to the problem of theodicy. I do not object if this book should be read as such an attempt, only with the proviso that it is taken as a "phenomenological theodicy," since it takes seriously phenomenology's claim of a consciousness both constituting time and being constituted by time. This claim has some important implications for the "ethical" self—or, rather, for the ways in which such a self informed by time—might come to understand anew the old "problem of evil." It is these implications that this book explores.

Acknowledgments

I am grateful to the President of the College of Holy Cross, Philip Boroughs, S.J., and to the Dean of the College, Dr. Margaret Freije, for granting me a one-year leave of absence during the academic year 2014–15 during which I was able to bring the present book to completion. I am thankful to all my colleagues on Mount St. James and beyond, especially to Professors William Desmond, Mark Freeman, John Gavin, S.J., David Goodman, Brian Becker, Christoph Schmidt, and Giulio Maspero for the "cross-pollination" among disciplines that made this book fertile in so many ways. Many questions behind the themes of the present book originated in my courses, and for this gift I am thankful to all my students at Holy Cross over the years. For the suggestion for the artwork included here, I would like to thank Andrei (Taake) Botez. I would also like to acknowledge a generous grant by the John Templeton Foundation which enabled me to present some of the material included in this book as a series of lectuers in Moscow. I am thankful to Fr. Vladimir Schmaliy and Andrey Shishkov for their hospitality while in the Russian capital.

I would also like to thank my editors at Bloomsbury Academic, Liza Thompson and Frankie Mace, as well as Juliet Gardner for their enthusiasm and professionalism in seeing this work to publication. Finally, for his unwavering support of my work, I would like to thank Dr. George Pelayo.

Part One

Prolegomena to Any Future

1

The Cosmology of Movement and the Metaphysics of Time

The question of time cannot be thought apart from the category of movement or motion (*kinesis*)—this seems at least to be the consensus of ancient tradition. Thus, both Plato and Aristotle understand time as a sort of movement and Plotinus as well as Augustine follow suit.[1] Of what time is a movement is of no concern to us at this early point. It is enough to note that the problem of time is invariably addressed by the philosophical tradition in terms of movement. This consensus necessitates for our discussion an excursion to the problems of movement and its relation to time.

Movement (*kinesis*) finds an application at two distinct, yet interrelated, levels: that of world history, a cosmology of movement so to speak; and that of the human being, an anthropology, or psychology of movement.[2] In the first chapter of this prolegomena we seek to describe the cosmology of movement or, as it may be better phrased, the metaphysics of *kinesis*. In the following chapter we shall discuss the anthropology of movement in relation to a phenomenology of time. We shall not elaborate at the moment on how the very term "metaphysics" implies a temporal dimension (this *meta*, or after) which, even when taken merely as an indication of sequence (*after* Aristotle's books on Physics or *after* such knowledge of the physical, empirical world) implies, however elusively, a system of thought transformed by time: there is always a metamorphosis in metaphysics or because of it. If we raised this curious lexicographical point is only in order to underline the irony implicit in the resistance which Greek metaphysics exhibited toward movement, the difficulty with which it was faced in thinking about movement, and finally, its tendency to demote movement from a problem to something simply problematic, something that pertains to the world of change and becoming, from which the philosophers ought to flee (if we were to believe Socrates in the *Phaedo*) should they wish to enjoy that which is supposed to be really their proper object of contemplation, namely, the really real, being as such.

The metaphysics of movement, therefore, might be even taken as a contradiction in terms. Nevertheless, such metaphysics of movement—to the extent that it is possible, and I intend to show in the pages to follow not only that is possible but also necessary—involves a complex cluster as well as a broad spectrum of topics: the metaphysics of *kinesis* covers the problem of be-coming, of whether the world is eternal or not, of genesis and creation, of multiplicity, of individuality and of the will, human and divine.

It is also the question of history and, above all, of the enigma of time. In other words, *kinesis* is a nodal point where cosmology, anthropology, and theology converge.

Given this importance, to which we have just alluded in the most general terms, it is pertinent to try to situate the metaphysics of *kinesis* within the history of philosophy. To do so, I suggest looking at three thinkers who, even though located distantly from each other by great chronological intervals, share a common preoccupation with movement and, ultimately, with time. Looking back in time, I single out the figure of Anaxagoras of Clazomenae, one of those original thinkers whom our need for classification, already since Aristotle's first book of *Metaphysics*, groups under the rubric of Pre-Socratic philosophy. Looking ahead in time, we meet the idiosyncratic writings of Søren Kierkegaard, whose thought, as I hope to show here, and as others have already argued,[3] was particularly preoccupied with the topics of movement and becoming. Between Anaxagoras and Kierkegaard stands the work of St. Maximus the Confessor whose thought, although usually read almost exclusively with an appreciation for its significant inputs to the obscure theological debates of his times, makes a decisive contribution to the ontological vindication of *kinesis* with all the far-reaching implications that such a position had for Christian metaphysics which had operated hereto with the categories, assumptions, and limitations of Greek thought. For this reason, Fr. Polycarp Sherwood's assessment of St. Maximus's thought finds its sharpest articulation as a refutation of Origenism.[4] By Origenism, of course, we should understand not only the work and thought of Origen himself but also of any Christian who was still the prisoner of Greek metaphysics, that is, of Neo-Platonic and Platonizing thought.

I should make clear that I do not suggest that Maximus was directly influenced by Anaxagoras nor that Kierkegaard was a reader of Maximus. Even though there is evidence that Maximus was at least familiar with Anaxagorean thought, and although one could make the case that one of Kierkegaard's intentional aims was, like Maximus's, to free Christianity from the burden of Origenism, nevertheless, the connection I am exploring here between these three thinkers is one that belongs to the genealogy of ideas, to the affinity of minds, and not to the tedious facts of historicism.

My thesis is that Anaxagoras's cosmology anticipates Maximus's more elaborate treatment of *kinesis* in the same way that Kierkegaard's appropriation of *kinesis* in the life of the individual—that is, in the task of becoming oneself—is anticipated by Maximus's theology of the will. As Anaxagoras rebelled against the Eleatic conception of a static being, so Kierkegaard launched a "polemic against truth as knowledge"[5]—that is, against a closed speculative system that had excluded the movement of personal will and, therefore, of faith.

A. *Thinking moves*: from Anaxagoras to Maximus

That something is what is not (what it appears) or that it is not what seems to be was the conclusion at which both Heraclitus and Parmenides arrived, albeit in different ways. It is important that both philosophical positions were ultimately motivated by how each philosopher positioned themselves vis-à-vis the problem of *kinesis*. Philosophy at this early and crucial moment of time became a question of denying or affirming *kinesis*. In this sense one could make the claim that all of the so-called Pre-Socratic philosophy is a philosophy of movement. Indeed, this is the case inasmuch as each of the Pre-Socratic philosophers attempted to articulate an answer to the problem of be-coming: Thales saw the unity of many in one element or origin. Anaximander saw the dispersion of the one into multiplicity as both the crime and punishment of becoming; Anaxagoras, as we shall see, saw in becoming a process of perfection, and so on. Furthermore, the repercussions of either the Eleatic or Ionian school assumed a meaning that gives philosophy as a whole the voice which we still recognize today when she speaks. The problem of *kinesis* was much more than denying or affirming a simple statement, for philosophy was now saying this: *what is is not*. For Parmenides this meant that sensory, empirical evidence of movement and change was a lie. For Heraclitus, on the other hand, this meant that the sensory, empirical evidence of constancy and permanence was a lie. How either of them read this statement is not important as long as we allow ourselves to stand in front of this decisive moment where a disjuncture, the non-coincidence of itself with itself, between the "apparent" world and the "real" world was first pronounced. Suddenly, the world became twofold. *All* spirituality, whether Christian or Platonic, finds its root and origin, its very foundation in this bifurcation of the world at which philosophy arrived, or rather, at which philosophy led us, when it contemplated the problem of *kinesis*. One of the aims of the present work is to re-examine critically the very nature of this bifurcation with the aid of a phenomenology that promises to give us back the world whole again, restored from such theoretical fractures.

Parmenides' argumentation in his *Way of Truth* seemed to have decided the question of *kinesis* once and for all. At the heart of that argumentation lies the principle of identity—the Parmenidean *is* (being) simply and most emphatically is—and that's pretty much all. Parmenides' thought reaches its highest moment in the articulation of the briefest utterance: ἔστι γὰρ εἶναι—*that which is* (what is usually translated as "being") *is*.[6] The implication of this flash of light is to illuminate the entire region of being. Parmenides should be credited for being the first to take note of the far-reaching consequences of this thought: what is, he says,

> (. . .) neither came to be nor could pass away,
> [it is] simple, immovable and without end.
> Nor was it ever, nor will it ever be; for now *it is*,
> all at once, one, and continuous.
> For what kind of origin should you search for it?
> How and where from could it ever increase?
> I shall not let you say nor think that it came from what is not;
> for one can neither think nor say that *what is* is not.[7]

Thus it becomes immediately clear that one could not walk *The Way of Truth*—that this is a way that leads nowhere, for there is nowhere to go, nor could anything ever move. Movement is not only impossible (Zeno's famous paradoxes of the impossibility of movement offered the layperson's version of Parmenidean ontology) but also erratic: it wanders away from the truth and thus into error.

Anaxagoras's response to Parmenides' description of a being in *rigor mortis* was nothing short of a stroke of genius. Parmenides had himself identified being and thinking (τὸ γὰρ αὐτὸ νοεῖν ἐστίν τε καὶ εἶναι[8])—after all, it was by means of his brilliant argumentation that he had reached the asphyxiating summit of the immovable One. And it was this very "reaching" that Anaxagoras had noticed: Parmenides denies movement *by moving* from one position to the next, from premise to conclusion. *Thinking moves*: this was Anaxagoras's way of saying *solvitur ambulando*.[9] And it moves not just itself, as in discursive reasoning, but thinking moves the world. Thus, this corrective to Eleatic metaphysics becomes the cornerstone of Anaxagorean cosmology where the *nous*, the mind, becomes a creative force of be-coming and of in-forming by being precisely the principle of *kinesis*.[10]

Subject to the well-known limitations of Greek metaphysics, Anaxagoras's vision of *genesis* is not a radical one—that is to say, it is not a becoming which presupposes nothing other than nothing itself (*ex nihilo*). The idea of creation, properly understood, does not fall within the scope of his system. Nevertheless, his brilliant cosmogony provides a description to the question of the world's transformation from chaos to cosmos. His ontology is inexorably connected with such a "cosmetic" understanding of the world, where τὸ καλόν (the beautiful) is, quite literally, *moving*: for it propels the world toward its future perfection. Here ontology meets teleology under the auspices of aesthetics as we find them again in certain passages of Plato, in the Greek Fathers, including St. Maximus and, finally, one last time, albeit in a much weaken form, in Kant's *Third Critique*. We shall have the opportunity to return to this connection between teleology and aesthetics from a different angle and in greater detail later on in this text.

This is at least what the attentive reader intimates from the celebrated passage in Plato's *Phaedo*, when Socrates offers, by a way of a long answer to Cebes' question after "the cause of generation and corruption" (95e), a surprising autobiographical account of his philosophical beginnings:

> One day I heard someone reading, as he said, from a book of Anaxagoras, and saying that it is Mind [*nous*] that directs [ὁ διακοσμῶν] and is the cause [or reason] of everything. I was delighted with this cause and it seemed to me good, in a way, that Mind should be the cause of all. I thought that if this were so, the directing Mind would direct everything [κοσμοῦντα πάντα κοσμεῖν] and arrange each thing in the way that was best. If then one wishes to know the cause [or reason] of each thing, why it comes to be or perishes or exists, one had to find what was the best way for it to be, or to be acted upon, or to act. One these premises then it befitted a man to investigate only, about this and other things, what is best. (...) As I reflected on this subject I was glad to think that I had found in Anaxagoras a teacher about the cause of things [τῆς αἰτίας περὶ τῶν ὄντων] after my own heart (...) Once he had given the best for each as the cause for each and the general cause of all, I thought he would go on to explain the common good for all, and I would not have exchanged my hopes for a fortune. I eagerly acquired his books and read them as quickly as I could in order to know the best and the worst as soon as possible.[11]

Socrates understands that Anaxagoras had offered two revolutionary answers. To the fundamental ontological question, "*why* is there something at all?" he had replied, "Because it is best": this means it is best for it and for the best's sake. To the question "*how* does something come to be?"—which now can be translated as "how does something become what is best for it," Anaxagoras had replied, "by being moved." Let us hear once more Socrates' interpretation: "If then one wishes to know the cause [or reason] of each thing, why it comes to be or perishes or exists, one had to find what was the best way for it to be, or to be acted upon, or to act." Let us, then, follow Socrates' advice and ask what state of being was considered by Anaxagoras as being the best for the world? Remember that Anaxagoras's concern was to offer an account of the world's transformation from chaos to cosmos. It is, therefore, of the utmost importance to note that "the best" as cause and reason of the world's existence was not, nor could have been, the *original* state of the world, but only a *future* state which, as future was still to come and the occasion of a be-coming, namely, of the world's movement toward it. At every given moment, therefore, the world exists because it is moved to a perfection ("the best") and by this perfection (the Nous). This perfection, τὸ τέλειον, cannot be but a feature of the end, τὸ τέλος.

Thinking moves, but this movement is not random nor spasmodic; rather, it is purposeful and all the more so as it is exercised by *nous*—that is, as far as it is a *intelligible* movement. It is, therefore, always a movement-toward, a movement directed to an end, for otherwise the inception of movement would have remained unaccountable and the world would have been unknown.[12] Contrary to Heraclitean flux, Anaxagorean *kinesis* is a movement with a beginning and an end—and above all a movement with a direction. Nevertheless, insofar as it remains restricted to the natural order, it is a movement that operates only within necessity; a movement for which the concept of freedom has yet to emerge.[13] And it is a movement still underway, for if it had reached its goal, movement would have already ceased and history would have already ended. Anaxagoras calls this movement by a term that should be familiar to the student of theology: *περιχώρησις* [*perichōrēsis*]. It is such a reading of Anaxagoras's vision which a youthful Nietzsche finds compelling and to which he gives the following gloss, so much reminiscent of Maximian eschatology:

> The end can only be striven toward in an enormously long process; it cannot be created all at once by a mythological magic wand. When someday, at an infinitely remote time, it is accomplished, when all the likes are gathered together and the primal essences lie side by side, undivided and in beautiful order, when each tiny particle has found its companion and its home, when the great peace enters the world after the great divisions and splits of the substances and no more split or divided material is left—then *nous* shall return to its self-movement, no longer roaming the world, itself divided, at times into greater, at times into smaller masses, as vegetative or animal spirit, indwelling in alien materials. Meanwhile the task is not yet at an end, but the kind of motion that the *nous* has figured out, in order to accomplish its end, demonstrates marvelous efficiency, for by it the task is nearer completion with each passing moment.[14]

Here two related ideas emerge simultaneously but also for the first time: first, a natural etiology that is teleological—what will be named in Aristotle's system as final cause and entelechy; and secondly, the ontological vindication of *kinesis*, where movement is not a falling away from some primordial perfection, nor the result of the One's dispersion into multiplicity (cf., Plotinus), but rather, to use one of Heidegger's phrases, *the event of being*. From the Parmenidean being to the Anaxagorean being is, indeed, a long way.

Did Plato lose sight of Anaxagoras's insight? As the quoted passage from the *Phaedo* suggests, on the contrary. Albeit his expressed disappointment with Anaxagoras, Plato's Socrates intentionally inscribes his own philosophical project within the Anaxagorean context, inviting us to think the one in light of the other. This becomes all the more apparent when Socrates suggests that his own philosophical methodology was precisely the attempt to construct what was the second best—the famous "second sailing" (99d)—to the high expectations which Anaxagoras's philosophy had raised for him.

This is nothing else than the theory of the *logoi*: ἐν τοῖς λόγοις σκοπούμενον τὰ ὄντα (100a). It is a common mistake to take Socrates' turn to the *logoi* as signaling the abandonment of his youthful inquisitions into the nature of things by becoming instead the ethical interlocutor interested in such things as the definition of virtue, etc.[15] In fact, there is a connection between Socrates the naturalist disciple of Anaxagoras and Socrates the ethicist: namely, *kinesis*, the desire to affirm the reality of change which both Parmenides *and* Heraclitus had denied.[16] The metaphor of the "second sailing" suggests a change of the *means* by which he intended to reach his destination, not a change of the destination itself. Socrates remains interested in the "cause of generation and corruption" of beings, only now this investigation is carried by means of the *logoi*—that is, by what is better known as the Platonic ideas or forms. If that theory became somewhat unintelligible in modern times it was precisely because it was understood outside of the Anaxagorean context that we have described. That is, the relationship between the forms and their participants was seen as static, operating, as if it were, on a vertical axis. But to deny the dynamism of this relationship is to leave it unexplained. Thus, Socrates insistence that "the beautiful becomes beautiful by the beautiful" (100d) sounds like an oracular riddle. What is missing in making sense of such a statement is the way one understands (or, alternatively, fails to understand) the form's capacity for production—literally for producing, i.e., for bringing forward, what it in itself possesses preeminently. If the Platonic forms are some mysterious, outworldly entities, their fecundity in making things the kind of things they are remains inexplicable—or else we are forced to come up with some magical explanation of "participation." On the other hand, if the forms are understood as perfections—that is, as both formal and final causes in the sense that God is portrayed in St. Thomas's Third Way—then participation is both the attraction which the form exercises on the thing by "moving" it (that is, by moving it to achieve its purpose) as well as the thing's striving to become more and more like its future perfection.

In short, what is missing from the explanation of the so-called theory of forms is the dimension of *time*, or more properly, the element of *time difference* between the two poles of participation or (trans-) formation. What forms, the idea, and what is formed are never contemporaneous. In a sense, what makes a thing the kind of thing it is is nothing else than the thing itself as seen in its future perfection.

The Platonic *logoi*, through the trajectory of the Dionysian corpus (in which they are baptized in the waters of Christian theology by becoming God's volitions or his θελήματα[17]) lead us to St. Maximus and in particular to the Maximian elaboration of the *λόγοι τῶν ὄντων*.[18] There is much more here than the seemingly coincidental borrowing of the Socratic *logoi*. In fact, St. Maximus in explaining a passage from St. Gregory Nazianzen's oration *On Love for the Poor* (*Ambiguum* 7) confirms the Anaxagorean provenance of the Platonic *logoi* for he identifies them with the Anaxagorean *moira* (μοῖρα).

> It follows, then, that each of the intellective and rational being, whether angels or men, insofar as it has been created in accordance with the *logos* that exists in and *with God*, is and is called a "portion of God," [μοῖρα Θεοῦ] precisely because of that logos, which as we said, preexists in God.[19]

Μοῖρα is unambiguously part of the main Anaxagorean vocabulary (occurring some seven times in the few lines of the preserved fragments). Its employment by Gregory Nazianzen is not surprising for there is textual evidence of other Anaxagorean references in his work.[20] Speaking of his friendship with Basil, Gregory uses the Anaxagorean language of "everything in everything."[21] The anonymous scholiast finds the opportunity to explain briefly Anaxagoras's views (see fragment B10). On another occasion, Elias of Creta commenting on Gregory's *Oration 31* identifies the Anaxagorean *nous* with God's *Logos* through which the world was created.[22] I cannot prove that Maximus had read Anaxagoras's work, but it is clear that he knew intimately Gregory's work and, therefore, I cannot deny him what other commentators of Gregory's orations have demonstrated. What remains uncontested is the fact that in Maximus's thought on *kinesis* we find a worthy successor of Anaxagoras's cosmology.

B. *A faith that moves mountains*: from Maximus to Kierkegaard

I propose now to examine the importance of the concept of *kinesis* for Kierkegaard's thought,[23] by paying close attention to the relevant passages from three works that are connected by their common treatment of *kinesis*, namely the *Philosophical Fragments*, *The Concept of Anxiety*, and *Repetition*.[24] I would like to suggest that Kierkegaard's treatment of this concept constitutes a largely unprecedented development in the history of philosophy, insofar as it re-evaluates movement/motion and, by extension, time and history as positive ontological categories against a prevalent model that favored immobility and immutability—the latter usually being credited to the Plotinian (and ultimately Platonic) triptych of *stasis-kinesis-genesis* where movement (*kinesis*) is only a falling away from *stasis*. Furthermore, what is of particular interest in Kierkegaard's refutation of that metaphysical hierarchy is that he is not treating the problem at hand abstractly, but rather through the concrete prism of salvific history, coming thus in confrontation with what can be called the lingering specter of Origenism over Christianity. *In fact, Kierkegaard's contribution on the subject of motion can be fully understood only as a refutation of Origenism.* In making this claim, I would like to compare Kierkegaard's treatment of the concept of *kinesis* with the thought of St. Maximus the Confessor who undertook, as demonstrated above, a similar refutation of Origenism by giving priority to *kinesis* over *stasis*, by reversing the aforementioned triad into *genesis-kinesis-stasis*.

Turning to Kierkegaard should also allow us to take notice of the differences between Maximus's theology of motion and Anaxagoras's cosmology of movement—despite the similarities that have been described already above. The chief difference between the Greek philosopher and the Christian theologian lies in the concept of *freedom*. Anaxagoras's worldview is ultimately aesthetical—note, for example, the emphasis on the nous as the decorating force of the cosmos—while Maximus's theology, even though sensitive and appreciative of cosmic beauty, sees in movement a force of *ethical* consideration as well. There is a certain automatism in the Anaxagorean march toward teleological perfection that disregards the volatile force of volition. On the other hand, for Maximus reaching the eschatological end is not something that either nature or history could ever achieve on their own, that is, apart from humanity's constant and free consent to God's plan. How this is so becomes clear by calling our attention to Kierkegaard's insistence on the philosophical importance of movement, to his understanding of the instant, and to his ethics of becoming.

Furthermore, Kierkegaard's critique of Hegel's speculative system is precisely motivated by his suspicion that Hegel's philosophy of motion, although it takes the form of an unfolding through a sequence of historical moments, remains ultimately merely aesthetic—the dance of the *Geist* through history, not much unlike the Anaxagorean *nous*. Kierkegaard seems to argue that Hegel's system falls short of grasping motion's full potential as an agent of ethical agency. What is missing is, in Kierkegaard's vocabulary, *interiority* which here means nothing more than taking into account the movements of the subject's will or, better yet, the will as movement. In his engagement with the monetheletic controversy,[25] Maximus was afforded the opportunity to develop a detailed anthropology of the human will (for him always based on and corresponding to Christology) that becomes complementary to the macroscopic role that movement plays in history[26] (see chapter 6, section III, and chapter 8 below). Therefore, any account that fails to pair Maximus's theology of the will with his cosmic understanding of motion is bound to remain incomplete.

1. Discovering Origen in Kierkegaard's works

References to Origen in Kierkegaard's works are sparing and cryptic. The only explicit reference to Origen in the published works is found in the *Concluding Unscientific Postscript*.[27] Given, however, the close affinity between the *Postscript* and the *Philosophical Fragments* (an affinity further reinforced by the author's note connecting the very argument by means of which Origen's name is invoked to the *Fragments*), we could trace Origen's specter through the pages of that latter work. And, indeed, we find, I believe, an allusion to Origen in a note at the beginning of the *Fragments* which criticizes the idea of the soul's pre-existence. Such a doctrine was a commonplace in pagan philosophy, yet there was only one Christian author closely associated with it—and later condemned for it—namely, Origen.[28] With regards, then, to this "Greek idea" (and notice that it would make little sense to accuse a Greek philosopher for holding a Greek idea, therefore, the author's target here could not have been Plato or any of the later Neoplatonic thinkers) Kierkegaard writes:

> . . . this Greek idea is repeated in ancient and modern speculation: an eternal creating, an eternal emanating from the Father, an eternal becoming of the deity, an eternal self-sacrifice, a past resurrection, a judgment over and done with. All these ideas are that Greek idea of recollection, although this is not always noticed (. . .) The contradiction of existence is explained by posting a "pre" as needed . . .[29]

The explicitly Christian context of the passage (creation, self-sacrifice, resurrection, judgment) helps us locate the reference to "ancient speculation" to a specific *time* in history. A note from Kierkegaard's *Journals and Papers* provides further evidence as to the *place* as well. There, the proposition of a pre-existence (of the Logos, of matter, of the soul, of evil) is more explicitly attributed to the "Alexandrian."[30] The same idea is repeated later on in the *Journals and Papers*, in the context of the same passage from the *Philosophical Fragments*: "[t]he contradictions of existence are explained by positing a 'pre' as needed (the Alexandrians) . . ."[31] The plural reference to the Alexandrians allows us the possibility of suggesting that Kierkegaard is thinking in this second note of Origen and Clement of Alexandria[32] and that the earlier reference to "the Alexandrian" must refer exclusively to Origen.

Of equal significance for our purposes is a lengthy note on the genealogy of the concept of the moment in Greek philosophy from *The Concept of Anxiety*: "This category," writes Kierkegaard, "is of utmost importance in maintaining the distinction between Christianity and pagan philosophy, as well as the equally pagan speculation in Christianity."[33] It is this "pagan speculation in Christianity" that becomes the target of both the *Philosophical Fragments* and *The Concept of Anxiety*. The expression "pagan speculation *in Christianity*" excludes Plotinus, but it allows us to include the Neoplatonic influence on a range of theological thinkers who write "in Christianity," from Origen to Schelling.

2. Kierkegaard's refutation of Origenism

Having established, however cryptically, an allusion that connects the idea of pre-existence with Origen in Kierkegaard's published works, it is now time to examine the place that this idea occupies in his thought.

Origenist metaphysics had, in effect, denied the reality of becoming, of this coming-into-existence in freedom and in the *moment* which Kierkegaard so passionately defended. Origen's position was captured in his vitiation of the beginning, especially an absolute beginning such that the creation of the world *ex nihilo* implied.[34] For Origen, not only did all souls pre-exist their primordial fall into the material realm but also—horrified as he was by the idea of an idle God that an absolute beginning of the creation in time seemed to suggest[35]—he had posited a continuous act of divine creation. That is, to the question "what was God doing before the creation of the world?"—a question that is now famous by means of St. Augustine's eleventh book of the *Confessions*, but which was first posed by the Alexandrian[36]—Origen had given an answer consistent with his metaphysical vision, that is, of a God who continually creates (and presumably destroys) a succession of worlds.[37] Therefore, even if *this* world is not eternal, as some of the ancients had believed, and, consequently, must have had a beginning (but, note well, *only in its material aspect*[38]), God's creative act, on the other hand, has no beginning.

The presupposition of Origen's system lies in the devaluation of *kinesis* (change, movement) and, as a consequence, it effects the vitiation of becoming insofar as an absolute becoming for the Greek theologies of Plotinus and the platonizing Fathers was tantamount with some kind of fall, a falling away from an initial perfection (represented by various ideas of Paradise) that was also, philosophically speaking, the *catastrophe* (the term originates precisely in this context as the "turning downward") from the One to many and from immutability to change. It is this Alexandrian (Origenistic and Plotinian) patrimony that Kierkegaard sought to refute by emphasizing the *moment*, precisely as *kinesis*, the change of becoming of the eternal into the temporal.[39] Thus, in an entry from his *Journals and Papers* which the editors of *The Concept of Anxiety* call "a significant notation," Kierkegaard writes: "The category to which I intend to trace everything, and which is also the category lying dormant in Greek Sophistry if one views it world-historically, is motion (κίνησις), which is perhaps one of the most difficult problems in philosophy."[40]

It is for this reason that the "companion piece"[41] to the *Fragments*, namely, *The Concept of Anxiety* takes hereditary sin as its point of departure, and, in doing so, it questions the Paradisiacal innocence and the concept of a prelapsarian perfection, both of which would have rendered the *moment* (or the becoming of history) harmful and disadvantageous, for movement would have signified a moving away from such an alleged perfection and a degeneration into imperfection. For the positing of such a time of perfection before history amounted to nothing else than what Origen had felt the need of doing when he introduced into his Christian dogmatics this *pre* of pre-existence, which, like a metaphysical *deus ex machina*, sought to explain away the paradox of historical existence. "When thought met with difficulties," Kierkegaard writes, "an expedient was seized upon. In order to explain at least something, a fantastic presupposition was introduced, the loss of which constituted the fall as the consequence."[42] This "fantastic presupposition" is precisely the idea of Paradise. "The understanding," he continues later in the same text, "talks fantastically about man's state prior to the fall, and, in the course of the small talk, the projected innocence is changed little by little into sinfulness, and so there it is."[43] But the current state of man is left all the same unexplained. "The more fantastically Adam was arrayed [in his pre-existence before the fall], the more inexplicable became the fact that he could sin and the more appalling became his sin."[44] We shall revisit the question of the fall in chapter 4 below.

Instead, Kierkegaard insisted on his paradoxical proposition that "the first sin is the sin"[45]—underscoring that, by failing to assume responsibility insofar as no man is tempted except by himself—for the serpent in Paradise, like the Sphinx at the gates of Thebes in *Oedipus Rex* [see. Chapter 5, section III], was the externalization of man's *concupiscentia*[46] (self-temptation, there is no other!)—one falls prey to the seduction of the seduction,[47] that is to the *ad infinitum* regression (pre-existence) of evil and, at the same time, the elimination of the decisiveness of the moment.

Similarly, the Origenist assumption of pre-existence makes "any point of departure *in time* … something accidental, a vanishing point, an occasion."[48] In the endless regression of eternity no point in time—nor, indeed, time itself—makes any difference. For, *sub specie aeternitatis*, the only vantage point for the Greek philosopher, time is indifferent:

> The temporal point of departure is nothing, because in the same moment I discover that I have known the truth from eternity without knowing it, in the same instant that moment is hidden in the eternal, assimilated into it in such a way that I, so to speak, still cannot find it even if I were to look for it, because there is no Here and no There, but only an *ubique et nusquam*. (…)
>
> If the situation is to be different [than the Greek paradigm represented by Socrates], then the moment in time must have such decisive significance that for no moment will I be able to forget it, neither in time nor in eternity, because the eternal, previously nonexistent, came into existence in the moment.[49]

And again,

> Let us recapitulate. If we do not assume the moment, then we go back to Socrates, and it was precisely from him that we wanted to take leave in order to discover something. If the moment is posited, the paradox is there, for in its most abbreviated form the paradox can be called the moment. Through the moment, the learner becomes untruth; the person … instead of self-knowledge he acquires the consciousness of sin …[50]

It is sin that is the difference between man and God[51] and it is that concept, the concept of sin, that Socrates, in spite of all his erudition on human nature, lacked[52]—and lacking it, the Greek philosophy remained innocent of the importance of time. As Socrates lacked the consciousness of sin, so Origen, by remaining within the Socratic, lacked the moment. It is under this light that Kierkegaard's "condemnation" of Origenism at the beginning of the *Philosophical Fragments* makes sense.

All of the above assumes a particular force in Kierkegaard's discussion of *objective* anxiety.[53] The very distinction between objective and subjective anxiety would make little sense if it was not clear that sin is as much a cosmic as it is an ethical category that pertains not only to the individual but to nature in general. In its cosmic consideration—applied, as if it were, on the macroscopic scale of the whole creation—sin's entanglement with time (for what else do we call now to mind if not history?) becomes extremely significant. The significance is underlined in Kierkegaard's gloss on Romans 8:19. "[I]nasmuch," he writes, "as one can speak of an eager longing, it follows as a matter of course that the creation is in a state of imperfection."[54]

What is this "longing" if not direction, movement, desire that stretches out (*epektasis*) to its ultimate goal at the end of time? Thus, movement becomes the indication of perfection insofar as this perfection has not been reached yet, *and* the means by which this perfection is carried out. The catastrophic aspect that Origen had attached to history and time is here reversed and annulled.[55]

3. Kierkegaard and Maximus

I would like now to draw very briefly and by a way of a conclusion, a comparison between Kierkegaard's refutation of Platonizing Christianity, as it has been sketched in the foregoing material, with a similar attempt undertaken by St. Maximus the Confessor. Maximus's main argument, according to the pioneering work of one scholar, was "grounded in the doctrine of motion."[56] One sees already here what prompted me to bring these two thinkers together. Maximus's positive re-evaluation of motion turned becoming, *genesis*, and, by extension, history into instruments of teleological perfection that, in his theological language, translates more specifically into salvation and deification. From the outset of his seventh *Ambiguum*, Maximus sets with the force of a metaphysical principle the fact that everything that moves does so as enabled by the power (δύναμιν) of its movement (κινήσεως) that is according to its desire (κατ'ἔφεσιν). This is a circumlocution to say that everything that is moved is moved by a desire seeking its fulfillment in the desideratum (ὀρεκτόν) that has been set by its nature. That history is still in progress is for Maximus, as it was for Anaxagoras before him and for Kierkegaard after him, evidence that things have not yet reached the fulfillment of their desire.[57] A little later, St. Maximus will clarify that to be created amounts to be moved and to be given being implies also the gift of motion. Thus, both the Origenist idea of a pre-existence outside time, as well as its corollary of the first man's existence in a paradisiacal perfection, are denied. Rather, humanity moves in steps that one of Maximus's famous triads defines as being, well-being, ever-well-being. The first and the last of this trinity of concepts is given only by God, since it *is* God: "[f]or the end of the motion of things that are moved is to rest within eternal well-being itself, just as their beginning was being itself, which is God, who is the giver of being and the bestower of the grace of well-being, for He is the *beginning and the end*."[58] The middle term, however, that is, "voluntary motion," depends upon us. In von Balthasar's words, motion "consists in allowing oneself to be carried by another in the depths of one's being and to be borne toward the ocean of God's rest."[59] And yet so much is at stake at this "allowing"—the whole drama of history is condensed here so much so that it might give the false impression of an almost effortless abandonment to its vagaries. On the contrary, "to be carried by another in the depths of one's being" consists in a life-long struggle punctuated with moments of dramatic anxiety, the kind which Kierkegaard often described in his works.

By this I do not suggest a Maximian influence on Kierkegaard. I simply draw attention to the curious convergence of these two critiques that stand apart in centuries, yet they are brought together by a common target, the platonizing Christianity of Origen, by a common solution, the vindication of *kinesis* and, perhaps, by a common orientation in thought that takes the incarnation seriously enough as to place Christ in its center.

2

The Anthropology of Movement and the Phenomenology of Time

The consensus of the philosophical tradition on time can be summarized, not without some simplification, as follows: time is in the mind, in the soul, in consciousness. Perhaps one could rephrase this in a slightly more sophisticated manner by saying that time is *of* the mind, *of* the soul, *of* consciousness. Thus, when Husserl writes his lectures *On the Phenomenology of the Consciousness of Internal Time*, he finds it necessary to begin his discussion by paying homage to Augustine's famous analysis of time in the *Confessions*. Augustine's definition of time as an extension of the soul had its own, earlier precedent, as we have seen in the foregoing section, in Anaxagoras's *nous* as the originator of movement that, in moving the primordial mixture of elements, transforms a world in a state of chaos into an orderly and unified whole: the cosmos. In a sense what Augustine does in the eleventh book of the *Confessions* is to translate an aspect of Pre-Socratic cosmology into Christian anthropology. Nevertheless, time as a phenomenon of the soul remains fundamentally unchanged in the history of philosophy up to Husserl, Heidegger, and Sartre. Yet, we are presented with certain aporiae: what does it mean to say that time is the outstretching of the soul into past, future, and present? Is the soul that *makes* time—*as if the soul somehow pre-existed time*—or is it rather time that *makes* the soul? Could one imagine the one without the other? And how are we to understand the outstretching, this *distension* of the soul[1] without committing the double error of making the soul a body, by assigning a material character to the soul and thinking of time as space, insofar as the soul is envisioned as some-*thing* that can be stretched out, as over some space? As Freud writes enigmatically in one of his posthumous notes: "*Psyche ist ausgedehnt, weiss nichts davon*."[2] The double ambiguity of knowing nothing is telling: if the soul knows nothing about it, about its own distention, about itself, how could I know if all my knowledge, and particularly the knowledge of my knowing, is by means of the soul? Is it even possible, then, to know time, insofar as time is *not* an object of perception but the condition of all perception? Or, better yet, how is time itself even possible?

As every student of Augustine's *Confessions* knows, these questions form the core of his investigation into the nature of time. "What, then, is time? If no one asks me, I know; if I want to explain it to someone who asks me, I do not know" (XI.14.17). Confronted with the enigma of time, one finds oneself divided—at the very least, this is a division between knowing and not-knowing. The question of time reveals this inner bifurcation that runs through the knowing subject itself [see Chapter 6, "Freedom

Beyond Freedom"]. At the same time and with respect to the same question I find myself both knowing and not knowing. Already here the distinction, if not the division, between reflective and unreflective consciousness is prefigured. It is important to take notice of this effect that time has on us as soon as the question "what is time?" is raised.

Of course, the perplexities of time are due to its nature: namely, that time, properly understood as the passing of time, as sequence and duration over time (a still time would either be a contradiction in terms or a category mistake, mistaken for eternity) can only be understood in relation to a past and a future, or, otherwise put, to what is *not*-anymore and what is *not*-yet, that is, it cannot be understood without reference to *nothing*.

> Now, what about those two times, past and future: in what sense do they have real being, if the past no longer exists and the future does not exist yet? As for present time, if that were always present and never slipped away into the past, it would not be time at all; it would be eternity. If, therefore, the present's only claim to be called "time" is that it is slipping away into the past, how can we assert that this thing *is*, when its only title to being is that it will soon cease to be? In other words, we cannot really say that time exists, except because it tends to non-being.
>
> XI.14.17

Here Augustine comes as close as one could to the mystery of time: what we know of time is that which cannot be known, for time presents itself precisely as a never-ending absence-*ing*. Any meditation on time places us face-to-face with the problem of absence. How to account for the presence of an absence?

It is rather embarrassing that both the simple man in the street and the educated man in the lab agree on this answer: my remembering of these words will be facilitated and carried on by a process that involves my brain's neurons. We should not concern ourselves with terminology at this point—synapses and all that—what matters is that either way (the way of science or the way of everyday naiveté) everything takes place "up here": in the brain.

This illusion of consciousness's immanetism—the belief, in other words, that every act of cognition takes place within the confines of the brain—arises precisely from the difficulty we have just named: the presence of an absence. Immanetism fakes what it cannot account for: absence. Since one cannot explain how consciousness does not need its objects present-at-hand in order to be a consciousness *of* them, the only other possible alternative, *in the absence of absence*, seemed to be the presence of the world with its objects *in* consciousness. A whole philosophical difference is played out in these grammatical propositions: where the "of" seemed inexplicable then the "in" had to be supplanted.

The problem with the mind's immanence is that it creates a dualism (of which philosophy is familiar since Descartes) between an outside (the things out there) and an inside (the world of thought in my head). Dualism not only leaves its own workings shrouded in mystery but it creates additional problems, three of which can be named as follows:

1. If mind and body are two distinct realities I am immediately presented with the problem of their correspondence: is there any relation of my perception of the things with the things-themselves? The nightmare of a Matrix is only possible from within a dualistic point of view.
2. Given the, at best ambiguous, correspondence between mind and world, truth itself becomes dubious or relative.
3. Finally, the possibility of a meaningful intersubjectivity, my relation with the Other, becomes irremediably wounded.

It is indicative of the primitivism of our thought the fact that we cannot think of consciousness but as a thing (a *concretum*)[3] or the property of a thing and, therefore, once unable to locate it or produce it, we feel forced to deny it.

1. The present absence: consciousness as intensity through the analysis of remembering

To assume that memory remembers by means of images, thereby every remembering becomes a form of re-presentation, will confront us with the very same problems St. Augustine was forced to face in Book X of the *Confessions*. These problems can be summarized as the conundrum of "imagining the unimaginable" since one has to account for the difficulty of ascribing an image to such memories which cannot, as we would say today, be "visualized": that is memories of emotions and feelings (X.14.21–2), or the memories of a piece of innate knowledge that presumably is not mediated by any image (X.11.18–12.19). When I think, for example, of freewill, do I recall the concept of freewill or freewill's image? And what would the image of a concept be like if not a concept itself? Is, then, the recollection of a concept mediated by the concept of a concept? And would it not such mediation required always new and seemingly endless intermediaries? What, then, about memory itself and forgetfulness? "Does this mean that memory is present to itself through its image, and not in itself?" There St. Augustine seems to come to the root of these paralogisms: forgetfulness is an absence—to say that I think of an absence by means of a presence (an image presented in memory) is to annul the very thing I am trying to think and to explain it by means of representation is to do away with what stands in need of explanation.

Let us continue briefly along the same line of inquiry by asking: Is consciousness the consciousness of someone? Is there, in other words, a *subject* of consciousness? And were we to say that, how we would have been able to distinguish between the two (the subject and its consciousness) in however a formal way? Would perhaps such a subject predate or precede its own consciousness with all the absurdities and difficulties that this might entail, such as of an unconscious subject that in some inexplicable way *becomes* conscious or obtains a consciousness as its *property*? Phenomenology has made clear that there is no room for such a subject in the life of consciousness;[4] that what one might continue calling a "subject" cannot be anything else than an embodied consciousness (and there can be no consciousness that is not embodied).[5]

The I does not *have* memories—that would still imply the mediation of the image—but it is that memory when it remembers, the image when it imagines, the thought when it thinks. So St. Augustine rightly writes "the person who remembers is myself; I am my mind."[6] There can be no distinction, therefore, between the memory of these words and I: while remembering myself writing these words, it is not I that am *having*, or the "I" that is having, that memory but rather I *am*, or the "I" is at that moment, that memory. The crucial clarification here becomes whether I am *only* my memory "and nothing more." We tend to assume that remembering is an act of being and not the whole of a being that remembers, that is, the remembering being. It is such an assumption that makes it necessary to search for a certain interiority into which the subject withdraws—itself or its other activities—while given to that memory, without losing its whole. However, is it not that whole already given in the remembering of that specific memory? Is it not, in other words, the case that in every act of consciousness the *whole* consciousness is invested without reservations, since any reservation would have been part of the investment? And likewise, therefore, for perceptions, imaginations, judgments, anticipations and all the other kinds of intentionality that make up the life of consciousness, our life. And yet, we admit that this picture, in its bold vision and elegant economy, does not satisfy us entirely. We feel that something more should be assumed and presupposed if we are to speak precisely as we did a moment ago of "*our* life." The question, then, becomes that of the principle that unifies the various intentionalities and guarantees the continuity of the self. The answer here is once again prefigured in St. Augustine's treatment of time in Book XI of the *Confessions* and it is telling that it is precisely to that text that Husserl returns in search of an answer to the unity and continuity of consciousness.[7]

Unity and continuity: to bring under the one, to appeal to the one. Two competing metaphors present themselves as distinct paradigms for the understanding of consciousness: is consciousness the operation of a transcendent ego which, like a spectator in theater or in the games, observes and, most importantly, by its observation unities what appears to it and what appeals to it for its coherence? Or is consciousness rather like the actor or the athlete of being, playing what it is and being what it plays?

Perhaps what is at stake in our questioning would become clearer if we transpose the same inquiry into a higher key. Let us ask then what constitutes the continuity of history? Speaking of personal history, the history of a particular individual, one assumes the existence of an I that endures through time and is thus the guardian of one's memories and by extension the guardian of one's own identity—the I as the archivist of my past. If, however, the discussion is that of trans-individual history, say, the history of a nation or of a people, or even the history of humanity as such, in the absence of such a universal I (besides, *who* or *what* could such an I have been or, indeed, how could it have been an I?) one substitutes a series of individual "I"s in succession. Our question is legitimized by establishing an analogical correspondence between the unity in past, present, and future of an individual act, such as the recitation of a poem; the unity in past, present, and past of a complete human life; and the unity in past, present, and past of human history itself. This parallelism was, in fact, suggested by St. Augustine when he wrote:

> What is true of the poem as a whole is true equally of its individual stanzas and syllables. The same is true of the whole long performance, in which this poem may be a single item. The same thing happens in the entirety of a person's life, of which all his actions are parts; and the same in the entire sweep of human history, the parts of which are individual human lives.[8]

What the parts of a poem are to the poem as a whole is what the deeds and actions of a person's life are to his or her life as a whole; and what a person's deeds and actions are to that person's life as whole is the lives of every human being to the macroscopic span of human history—and this is a unity brought about by time and over time.

The juxtaposition of these two levels of history reveals a problem: in the case of world history one dispenses with the necessity of one, individual, permanent I without being bothered that the fragmentary plurality of individual "I"s—all equivocal among themselves—could not in fact account for what we sought to establish in the first place, namely, continuity. These individual "I"s are born and die like moments: preceded by nothing and followed by nothing. How could they establish continuity insofar as nothing from the one survives in the next? And yet, there is history, there is the memory of a past that exceeds me and every synchronous I by far. The past of my hometown, for example, two thousand years ago, is still shaping and forming my experience today. Thus, *we know* that continuity is not the byproduct of synchronicity; in fact, synchronicity, by its nature, knows nothing of continuity and the experience of history.

Why do we, then, still demand of the particular and individual I to be such an allegedly synchronous entity? Because, as Kant rightly said, each of our experience must be accompanied by an I.[9] We begin to see now, however, that this I cannot be the selfsame, identical I throughout, unless I am prepared to admit that one has only one experience given in a *nunc stans*. In other words, it makes no sense to assume one single enduring I (synchronicity) in order to account for the unity of the multiplicity of my experiences, for what one has to sacrifice then is nothing less than the same multiplicity of experience, the unity of which one seeks to safeguard.[10] At the very minimum, we must admit that the I finds itself at different chronological moments, that this I has time and is in time. Admitting, however, the diachronicity of the I is to admit already too much: there can be no one I but, like in our example of world history, a succession of "I"s that transmits to its next not only its own content but that of all its predecessors.

Incidentally, such was also Brentano's theory of the origin of time, which Husserl criticizes precisely as failing to take the past as that which has passed (i.e., as an absence) into account. Instead, Brentano supplanted an image for the thing remembered but then, as Husserl rightly observes: "where do we get the idea of the past?"[11] Husserl's answer to this problem was that every act of perception is surrounded, as if by a halo, by absences. In order to see what is in front of me, in order to see it as that which it is, it is not enough to see what I see but it is equally necessary to "see" what I do *not* see. When I look at the facade of a building, it is not enough to see the side of the building in front of me, it is equally important that the other sides of the building which are not within my visual field co-present themselves together with the one that presents itself to my vision—but they co-present themselves not as present but precisely as absent sides. I can never see something *all at once*: in a moment of pure, unadulterated presence. The perception of any physical thing, that is, any perception in space and time, "involves a certain *inadequacy*."[12] By this Husserl means that what I see is always necessarily partial, for I can never grasp what I see fully, from every single angle, in every possible way an object can show itself to me. This partiality, this imperfection endemic to perception itself, is a limitation necessitated by the limitations that are imposed on both me, as the perceiver, and the object of my perception, on account of our respective embodiments. However, the same imperfection perpetuates a series of inexhaustible possible perceptions "which can always be continued" and "which are never completed."[13] This characteristic alone is enough to become the criterion for distinguishing between two kinds of beings: being as a physical thing, and being as an act of consciousness (for example, the distinction between the perception of a thing and the consciousness of that perception). The former is always given through a multiplicity of adumbrations, the latter can never be perceived adumbrated.

This very rudimentary sketch of one of the fundamental positions of the phenomenological method should suffice in making clear that phenomenology was prepared to account for the presence of an absence, to pay equal attention to what is absent as philosophy had in the past done for the present and, therefore, it was natural that time, as the process of absence-*ing*, should become a proper subject-matter for phenomenological analysis.

The phenomenological analysis of internal time reveals time as the perichoretic intertwining of three ecstasies: impression, protention and retention. It is important to note that in what we call the "now" all three temporalities are to be found: there is a present-present, and a present-past and a present-future.[14] And the present-present, in turn, springs its own branches of protention and retention. This seemingly endless process affords us with a much more nuanced conception of temporality than that of linear succession of "nows."[15]

Thus, the naiveté of immanetism is averted. The memories I have are not stored somewhere in my mind to be recalled or put aside at my will as one picks photographs from a drawer. Neither the I, nor its "contents," have an essential permanence: they are not solids but moments. Consciousness has no depth, no interiority to which it could withdraw. It can assume "weight" or "force"—that is intensity but, as Bergson has already shown,[16] has no extensity. Hence, the existentialist manifesto of existence's priority over essence and everything that this axiom implies.

2. Waiting memory: time as intensity through the analysis of boredom and waiting

In previewing my day, as in these moments when I lie awake but still in bed, my life takes the form of a series of "things to be done" (*agenda*). Some of these "things" are already there as part of my living existence (e.g., lunch)—I do not normally think of them, unless they rise to an unusual event, e.g., lunch *with* a friend, but they provide my schedule with its frame. Then, there are some tasks that need to be performed as part of my professional life, e.g., classes to be taught, meetings to be attended, etc. Through them, my life is carried on. They present themselves as stepping-stones that can take me across—to the same point where, lying in bed and about to fall asleep I might repeat the morning review of my day. What we call life is a series of intervals from sleep to sleep.

> In the morning, when the hour strikes at which I am accustomed to rise, I might receive this impression σὺν ὅλῃ τῇ ψυχῇ, as Plato says; I might let it blend with the confused mass of impressions which fill my mind; perhaps in that case it would not determine me to act. But generally this impression, instead of disturbing my whole consciousness like a stone which falls into the water of a pond, merely stirs up an idea which is, so to speak, solidified on the surface, the idea of rising and attending to my usual occupations. This impression and this idea have in the end become tied up with one another, so that the act follows the impression without the self interfering with it. In this instance I am a conscious automaton, and I am so because I have everything to gain by being so. It will be found that the majority of our daily actions are performed in this way and that, owing to the solidification in memory of such and such sensations, feelings, or ideas, impressions from the outside call forth movements on our part which, though conscious and even intelligent, have many points of resemblance with reflex acts.[17]

In the meantime, one does things—so much so that in the event that one has nothing to do, one must find things to do.

"Nothing to do": this phrase presents us with an important clue in our analysis. "Nothing to do" reveals that in between these "to dos" lies precisely nothing, that the stepping-stones of daily tasks, and goals, and events take us through the dark waters of nothing. Yes, in the unlike eventuality when one is left with "nothing to do" one must find, invent, things to do, lest he is swallowed up by nothing.

When is one left with nothing to do? Is it even possible that one can be left with nothing to do? It is not the possibilities one lacks, and perhaps not even the actual tasks—for I feel that I have nothing to do when I have a lot to do. It is rather the motivation that one finds lacking. I am "stepping" on my first task, which in all likelihood I have done already a countless times and, suddenly, the next step seems impossible or better yet meaningless. Impossible, insofar as it involves an insurmountable effort. Meaningless, for I cannot see any reason why I should carry on with this futile game. Yes, I can step on the next stone, which means I can hold my next appointment, I can perform my next task but so what? What is the point? Sometimes or for some people this is the point: avoiding the pointless and therefore, one throws oneself in the faithful execution of steps. For one such as this, however, who has already caught this vision of pointlessness, proceeding is indeed and physically impossible. The futility of what he called "a game" is more than a thought. In fact, even before rising to mind as a thought, futility paralyzes one's members with what Levinas has called the stiffness and the numbness of fatigue: "Fatigue—even and above all, the fatigue that is unthinkingly termed physical—presents itself first as a stiffening, a numbness, a way of curling up into oneself."[18]

But why should I lack the motivation, especially since this is not a motivation for this or that but eventually motivation for life itself, that is, the lack of a *conatus essendi*? Is it not that futile game that we described above the masquerade of desire that knows no end? Is this continuous succession of goals and tasks, of things to do and to be done, not propelled by an insatiable desire that desires only its own continuation, its empty desiring? How then could one lack motivation? How could one lack the lack that desire is? A double lack.

Fatigue demonstrates not the lack of motivation but rather a suspension or better yet an *obsession*. Desire is not mitigated but rather frustrated and this frustration signals an intensification of desire. In fatigue desire places itself on hold. It waits and it is waiting that explains what appeared earlier as a "lack of motivation," a "refusal to go on." While waiting, there is nothing to do, perhaps because one is already consumed in doing something, namely, wait. While I am waiting for my train to arrive I cannot seriously engage with anything because a) my primary engagement is precisely the waiting-for-the-train-to-arrive and therefore anything else (e.g., reading a book, smoking a cigarette) cannot but be only a way of "measuring" the time of waiting, i.e., a form of parergon to waiting itself; and b) the arrival of the train can interrupt these other activities rendering them thus pointless. Therefore, as long as I wait I cannot do anything, there is nothing to be done.

The point in this rather commonplace analysis that needs to be stressed is the following: if there is nothing to be done while waiting it is neither because I cannot find something to do, as if I were lacking the means or the ideas, nor because I do not want to do anything else, as if I were lacking the desire. I am neither imprisoned nor bored. In fact, there might be many things that I would like to do, however as long as I am waiting I can do nothing.

To say, as we have just done, that "as long as I am waiting there is nothing to be done" is to say precisely the opposite from what Samuel Beckett showed us on stage in his play *Waiting for Godot*. The opening words of that famous play are exactly these: "Nothing to be done." But whereas Vladimir and Estragon wait because there is nothing to be done, we, on the contrary, have nothing to do, *because we wait*. The difference is decisive. One would be justified, I believe, in saying that what Beckett put on stage was a grotesque exaggeration of what Heidegger had predicted, in one of his dark prophesies concerning our scientific age, as the "hidden goal toward which all of this and much more rushes, without having the slightest hint of it—and without being able in the slightest to have a hint—is *the state of boredom* ..."[19] For Heidegger, boredom is modernity's destiny, as much as it was, ironically, for Greek antiquity the primordial origin of man. We remind the reader that for someone who thinks from within the paradigm, and indeed the culmination, of Greek metaphysics, like Origen, history is the result of the boredom (κόρος) that the pre-existing souls experienced in a state where there is nothing to be done other than contemplating the Good.[20] Yet, we overlook the possibility that Beckett's *Waiting for Godot* might also be a very ancient story: the story of an endless (and thus pointless) repetition, of the infinite return of the same, both at the beginning as it is in its end. Heidegger's analysis of boredom belongs to such a conception of homogeneous time as cyclical *chronos*.[21] However, it is quite a different conception of time that is operative in our analysis of waiting insofar as waiting, even when one knows not what one waits for, or especially then, is open to what we can call an eschatological expectancy which is neither for "the beings as a whole" nor for Dasein's "resolute self-disclosure."[22]

Not everything, however, arrives with the precision of a train. Indeed, many events lack a schedule: one can anticipate them but not their arrival. Thus, one does not quite know what one waits for. There is an element of indeterminacy as to the object of one's expectation. One of course is free to give it different names and imagine it under different categories: it might be a better future; a phone call that would magically transform my dull life; the beloved that I have always dream of; perhaps death or the Messiah. Insofar as one can have anything to expect there is waiting. As long as I wait, there is nothing to be done. Every action as long as it can be seen in the horizon of a time suspended by waiting is pointless.

In another sense now, waiting itself is an activity. "[T]he existence of an existent," Levinas writes, "... is by essence an activity. An existent must be in act, even when it is inactive."[23] Thus, the expression "nothing to do" implies now that in waiting one indeed *does* nothing (where this "does" is understood in all its active sense). Think of the paradox of an "active inactivity" as the tension[24] expressed in the positioning of an athlete who is about to run—to return to one of our opening metaphors. The athlete has not yet begun running and he is not running. In positioning, he assumes the immobility of a statue, frozen, as it were, in the instant. And yet, all the subsequent activity, the energy that would be released once he leaves his mark, is gathered in the tension of his muscles. There is an *entasis* in waiting. I am borrowing this technical term from classical architecture, a term that usually denotes the convex curve of a column. The word "entasis" itself can be translated as *intensity*. To the question why the masters of old applied the device of *entasis* on their buildings—superficially understood as nothing more than an optical illusion—one would have to answer with reference to the anthropic ideal of classical architecture, where the proportions of the human body were applied to buildings, but also to extraordinary sensitivity of the ancients that understood perfection as including imperfection, or rather, they understood that there is no perfection without imperfection, and therefore created columns which, like the muscles of the body, bulge under the weight that they support. Waiting is such *entasis*, such intensity. But this is a paradox, perhaps greater than that of an "active inactivity." We understand waiting as a category inseparable from the time one waits for. There is no waiting without time. But also, we understand time as extensity: as the measure by which one calculates the duration of an activity. So, it takes me fifty minutes to go through one of my lectures and five minutes to walk from the classroom to my office. In talking in this way, we imagine time in terms of space.[25] In the second example, time is a correlate of my moving from point A (say, the classroom) to point B (my office). In the example of a lecture, time is the correlate of my covering the "distance" from the beginning of my presentation to its end. Here again, although more subtly, one thinks of time in terms of space, that is, in terms of extension. Although this conventional way of thinking of time cannot be dispensed, since it is part of our everyday language, the analysis of waiting has revealed another aspect of time, namely, temporality as *intensity*.

We noted above that when one waits for certain events there is a certain indeterminacy in both the form of the event (e.g., I am waiting for a better future but I do not know, in fact I cannot know, in what form that future might arrive) as well as its time (its "when"). What I am expecting does not always come with a predetermined schedule—in fact, I might not even be sure that it will come. Like Estragon and Vladimir, I am not sure whether it is today or tomorrow or in fact that it might have been yesterday that Godot was to come. In the absence, therefore, of a fixed point B, a terminus by which one might feel justified to give up one's waiting (for after that one *knows* that what one expected has either come or not), the time of my waiting lacks all extensity. The moment I am waiting for is not anymore di-stance but rather *in-stance*: it might be "any moment now," or worse, "already here." So *waiting becomes a pure intensity*—under which everything else and above all economic time collapses.

The phenomenological given of a consciousness as sheer spontaneity, as a moment, that derives its unity not from itself but rather from its objects so much so that one could dare to say that consciousness itself is not only a subject, or not primarily a subject, but also an object; that is, consciousness does not only constitute the world by its intentionality but is also, in turn, constituted by a counter-intentionality—all of this, described and explained in detail by Sartre and Levinas and Marion, present us with a subject without subjectivity defined by its passivity or affectivity.[26]

The foregoing analysis was necessary in order to arrive not once but twice to the phenomenon of intensity. I have shown how consciousness can be understood as intensity; but I have also tried to show that time assumes the same characteristic of intensity. The conclusion that I hope to draw at the end of this comparison should be fairly obvious now: consciousness is time. Time is the event of being—a being understood not in terms of substance but of becoming. This, however, does not mean that time "comes from the self": that would imply the untenable position of an uncreated self.

I believe, in agreement with Sartre[27] and Levinas,[28] that the instant that is the life of both consciousness and time constitutes a creation *ex nihilo* and, indeed, to such a degree that it requires a continuous creation.[29] This ever-continuous opening of the self to the Other that constitutes its "resurrection" is what Levinas calls in his *Diachrony and Representation* the "deformalization of time":

> It was important to me above all to speak in this study about how, in the human intrigue, past, future, and present are tied together in time, without this resulting from a simple degradation that the unity of the One could have ... undergone, dispersing itself, in the movement that since—or according to—Aristotle was acceded to time in its diachrony. In such a view the unity of time would lose itself in the flow of instants, to find itself again—without truly finding itself—in re-presentation, where the past gathers together instants by way of the memory's images, and the future by way of installments and promises. But I have sought for time as the deformalization of the most formal form that is, the unity of the *I think*. Deformalization is that with which Bergson, Rosenzweig, and Heidegger, each in his own way, have opened the problematic of modern thought, by starting from a concreteness "older" than the pure form of time ...[30]

3. The eschatological constitution of consciousness

Leaving aside for the moment the technicalities of this analysis, it would be useful to open up our discussion to a broader scope so that it will include some theological considerations, keeping in mind that the question of time in Augustine (as well as in Anaxagoras) is raised against such a theological background. I spoke above of the phenomenological understanding of time as the "perichoretic intertwining" of three ecstasies: impression (the present-now), protention (the not-yet-now) and retention (the not-anymore-now). The term *perichoresis* is a term that we first met in Anaxagoras's fragments (B 12) indicating the movement that propels the universe to its teleological perfection. It is the movement we know more concretely as *history*. Taken up by the Christian authors of the fourth century, the same term became a technical term that describes the indwelling of the three Persons of the Holy Trinity. Even though Augustine does not explicitly makes such a claim, it would be safe, I think, to see in the perichoretic intertwining of attention, expectation, and memory (XI.28.37) a *vestigium trinitatis*, that is, the trace of the Trinitarian God.[31] Such claim has the far-reaching implication of a reduction of temporality (in the sense of *re-ducere*, leading back to) to the heart of God's existence—an existence that moves away from the notion of a static being, but upholds both being and becoming and the coincidence of being with becoming.[32] Incidentally, this is also Husserl's insight: for what else is the "absolute flow," the *stehend-strömended Gegenwart*[33] which, for Husserl is the origin of time and thus somehow outside time, an unconscious consciousness,[34] if not the στάσις ἀεικίνητος (the ever-moving rest) that someone like Maximus the Confessor uses to describe the temporal mode of humanity's eschatological destiny?[35] Thus, time is not any more divorced from eternity, but rather finds its origin, and therefore its goodness, in it. At once close and yet distant from the definition that Plato gives of time in the *Timaeus* as the "moving image of eternity" (37c-e), time is not the byproduct of an immemorial fall, a catastrophe away from the stillness of the One, and thus a deplorable characteristic of a fallen humanity from which one could only seek and hope to escape. Time, in this theological reading that phenomenology has presented us with, is the means to participate in a process of perfection.

> Life is then called a way because each being that enters into life hastens toward its end. Just as those who are sleeping in ships are carried by the wind through its own force to the harbors, even though they themselves do not perceive it but the course hurries them on to the end, so we also, as the time of our lives flows on, are hurried along as if by a continuous and restless motion on the unheeded course of life, each one toward his proper end.[36]

It is precisely such a salvific understanding of time that St. Augustine illustrates in his biographical narrative of the *Confessions*, and it is this role that time has played in his journey from the garden of Thagaste to the garden of Ostia (or, reading his history writ large, from the garden of Eden to the eschatological garden of Paradise) that time as such needed to be thematized in book XI. It is time that unites the two seemingly disparate halves of the *Confessions*. For Heidegger, book XI "provides the essential 'inner jointure [*Fügung*] of the *Confessions* as a whole.'"[37] We shall return to this theme in great detail later on.

In a recent publication the argument was made that our brains contained, as part of their evolutionary trace, the notion of God and that, therefore, God was to be found in the mind.[38] For the phenomenologist, however, nothing can be found in the mind, for the mind, or better yet consciousness, is not a container but rather, in Husserl's terms, an *act* that takes place "out there," amidst the world. Nevertheless, the phenomenological analysis of consciousness reveals a similar position as the one circulated in periodicals of popularized science: it reveals the heteronomic origin of consciousness. The life of consciousness bespeaks of a constitution that comes not from itself but from another—indeed, from the Other.

In the following pages, I would like to suggest that consciousness's teleological capacity to perceive beauty, pleasure, and perfection suggests not only an orientation toward the "not-yet," and thus an opening to the eschatological, but also it reveals a constitution that could be rightly called *theo-logical*.

4. The three calls of givenness

In the first formulation of his own "broadening" of the phenomenological reduction, Jean-Luc Marion discovers a horizon more essential than, and thus anterior to, transcendental consciousness (Husserl) and being (Heidegger). What constitutes phenomena and, by extension, what constitutes me, as the recipient of these phenomena, is neither the intending character of the consciousness paired with the phenomenon's intuition, nor is it the opening of the Dasein to the nothingness of Being disclosed by anxiety and boredom, but rather the claim addressed to me by "the pure form of the call." Thus Marion writes "that which gives itself gives itself only to the one who gives himself over to the call and only in the pure form of a confirmation of the call, which is repeated because received."[39] Marion's discussion of the call is indebted to Heidegger's analysis of the character of conscience as a call that calls Dasein to itself, a call that "comes from me yet it calls from beyond me," as section 57 of *Being and Time* famously stated. Yet, Marion radicalizes Heidegger's analysis by emphasizing that the very receptivity of the call is constitutive of a subject without subjectivity for it is neither a being nor a consciousness.[40] It should be noted that the subject does not even exist prior to the call, for "giving himself over to the call" means, first and foremost, to "be given a self by the call." The self that gives himself over to the call does not have even himself; in order, then, to give himself over to the call he has to be given that self. In fact, this is not about a sequence, logical or chronological: the self is not *first* given in order to be later given up, but rather the self is given *as much* and *insofar as* it is given up. For the self too, or rather the self above all, must be given.

Indeed, the gifted (*l'adonné*) is called to existence as a response to a call (*l'interloqué*) that calls it to being. "Thus is born the gifted," writes Marion, "whom the call makes the successor to the 'subject,' as what receives itself entirely from what it receives."[41] The summon of the call, the resulting surprise, the call itself and its facticity—this fourfold of the phenomenology of givenness—imply a self given to oneself by an origin that precedes and predates it and, at the same time, the paradox of a self who, in receiving itself, precedes also and predates itself.

The call, exemplifying what is known as an inverse intentionality,[42] summons me to myself. It summons me, neither to a diluted subjectivity where no taste of particularity can be detected, nor to the ontological uniformity of one-size-fits-all, but rather to myself, that is, to the irreducible, irreplaceable specificity of my thisness. "The passage from the nominative [of the subject] to the objective cases (accusative, dative)," Marion notes, "inverts the hierarchy of the metaphysical categories."[43] But how? And what may be the implications of such an inversion? Is the project of Marion's reduction based only on a grammatical whim? Certainly not. What is at stake here is by far more radical than any Copernican revolution. Marion explains it in what, in our opinion, might be the most far-reaching claim of his phenomenology: "Individualized essence (*ousia prote*) no longer precedes relation (*pros ti*) and no longer excludes it from its ontic perfection. In contrast, relation here precedes individuality" and, as he adds a few sentences later, it results from it.[44]

Alluding to Dionysius's *Divine Names*, we could say that the self is like the name. A self must be given as a name is always given—we speak of somebody's "given name"—and never assumed by myself. I cannot name myself unless the Other first gives me my name, by calling me—my parents after my birth, the priest in my baptism, the abbot in my tonsure—thus giving my self to myself. In the absence of others there is neither name nor self. To be given a name indicates one's beginning, in my name I acknowledge that I am generated, derived and depended; the fact that I have a name by which the Other can call me implies that the Other has laid a claim over me, that I belong not to myself but to the Other from whom I received not only my name—my name, after all, is a constant confession of this debt—but also my self. A name is always given and therefore it can never be a proper name—for my name does not belong to me, not only insofar as it is given to me but also insofar as it has named others before me and it will name others after me. However, "in this way, the baptismal given name, the 'proper' name par excellence, results from a call (one calls me with the name of such a saint) because, more essentially, this name constitutes a call in itself—I would not be called simply *by* this name, but indeed *to* this name."[45]

Only God has no name for God has no beginning. Who was there before God in order to name Him? To give God a name would imply that oneself is prior or higher than God but such "God" would only be an idol, for he who names God creates "God," that is, he erects for himself an idol. God is strictly anonymous (as the *Divine Names* make clear) or, and this amounts to the same thing, polyonymous.[46] Marion's early critique of conceptual and metaphysical idolization (cf., *God Without Being* and *The Idol and Distance*) finds its complementary gesture in the critique of subjectivity, for the subject is the idol of (one)self.

It is precisely the death of such idolatry, the idolatry of the self-subsisting subject, that the triple immersion in the baptismal waters effects so as the new person who emerges from death can now receive not only a name but a new identity "in the name of the Father and of the Son and of the Holy Spirit" that is, an identity inscribed within a community, a personal identity, since it is given by the invocation of three Persons, given as a gift and not claimed as a possession. For "what do you have that you did not receive?" (1 Cor. 4:7).

After the call to biological existence and the call to ecclesial life there is finally one last call: "I tell you the truth, a time is coming and has now come when the dead will hear the voice of the Son of God and those who hear will live" (John 5:25). Here, then, lies the whole question of the call, for "the dead will hear the voice." How are the dead to hear the voice that calls them to everlasting life? How is the voice to penetrate the dead ears that hear nothing? To play with this paradox, we could say that only the dead ear hears; that unless one has become (like the) dead one would not hear the call. Indeed, receptivity to the call presupposes an unconditional passibility, such that one could compare it only to the absolute passivity of the dead. As in the *Mystical Theology* one ascends higher by leaving behind more and more of oneself, that is, of one's own categories, concepts and images as to arrive to the summit of *aphaeresis* naked of all conceptual armory, so here we shall hear the voice that calls to life once we have silenced our voice and have mortified the activities of the subject, once our intending consciousness ceases to search and returns upon itself by means of an inverse intentionality that allows it only to receive, to hear the call and by hearing it to live. The domain of the call, particularly of the divine call, extends from before birth (cf., Luke 1:13) to after death (cf., John 11:43; Luke 7:14) and thus proves itself of being unrestricted by what we might take to be life's absolute termini. Similarly, though, the one called, the one to whom the call is addressed is shown to be more than his being for, in a sense, he precedes his birth and survives his death. Thus the call has displayed the insufficiency of being or consciousness in counting as the ground of selfhood. The *interloqué* is "man without being" as the call that constitutes him comes ultimately from the "God without being."

The foregoing analysis might give to our discussion of the call a particular color that is not overall accurate. It might, in other words, give the false impression that the call is restricted to one category, that of religious or ethical phenomena; for instance, the "call of conscience" or vocation as a "calling." Against this assumption, we must emphasize that the call is above all a property of the visible, or better yet, the call becomes most noticeable as the visible. By this we mean not that everything that appears is the call, but rather that whatever appears (from everyday things, like chairs and books, to ideas, emotions and state of things) appears because it addresses us a call. A silent face in a café, a painting in a museum, an exam that I need to take, all of these appear by means of a certain call, a call to which I can respond in different ways. The call is not only what calls our attention but also what fails to do so, the unnoticed and the unnoticeable; the call is not only the pleasing or the interesting but also what one finds unpleasant or boring. Therefore, it makes all the more of a paradox to say that the call is the beautiful.

5. The promise of the beautiful

It is now time to elaborate more attentively on teleological character of the beautiful to which we alluded earlier on in our text. Of course, one knows, since Kant, that the beautiful is not to be identified with pleasure and von Balthasar does not hesitate to take even ugliness as a manifestation of God's glory (the unsurpassable example of such paradox is of course the cross). Yet, what does it mean that the call is the beautiful? Indeed, what else can the beautiful be than what calls? And, how else is one to understand the ability of the call to call if not by means of beauty?

Language tells us that much when it indicates that the derivation of "the beautiful" (*to kalon*) comes from the verb "to call" (*kaleo*, *kalein*). Naturally, if we understand beauty as symmetry or proportion, as harmony of color or sound it would be difficult, indeed impossible, to explain the catholicity of the beautiful as the call that calls through the visible, even when it is not a question of harmony or symmetry. These "scientific" explanations, as somewhat scornfully Socrates calls them in the *Phaedo*, are descriptive at best of the ways in which beauty is perceived, that is, they explain only the "mechanics" of the aesthetical phenomenon, but fail to answer why we call something beautiful or, worse, what beauty is in itself. Plato, therefore, rejects them as insufficient and confusing, Kant as threatening beauty's universality, and phenomenology as imposing limitations that are inadmissible within the reduction. The only two answers to the question of beauty that merit some consideration are those given to us by Plato in his *Phaedo* and Kant in his *Critique of Judgment*.

For Plato, "what is beautiful is beautiful by the beautiful" (100d, 7–8). Of course, such a statement is heavily in need of interpretation. One has learned to see in this answer Plato's so-called theory of forms. The beautiful, then, by which anything becomes beautiful is taken to be the form of beauty. This already implies that what makes something beautiful is not itself, i.e., it is not to be found in the thing itself, but rather comes from beyond, it is other than the thing that one perceives as beautiful. Surprisingly, Kant gives a very similar answer when he refuges to assign beauty as the property of a thing.[47] For him, too, beauty is external and a sign of exteriority. Both Plato and Kant seem to converge on another point: that beauty is teleological. It is needless to rehearse here the movements of Kantian teleology—suffice to say that it is solely the teleological character of the beautiful that maintains the coherence of an otherwise disparate *Critique*, divided, as it is, between aesthetic and teleological judgments. To see a similar notion in Plato's treatment of the beautiful we need to remind ourselves of the context within which he discusses beauty: it is the famous episode where Socrates gives a brief account of his philosophical autobiography and of his encounter with Anaxagoras's teleology in particular. Socrates believes that in Anaxagoras he has found the only tenable answer as to the cause of things, that is, *perfection* ("for if one wished to know the cause of each thing . . . one had only to find what was best for it" 97c). His later disillusionment with Anaxagoras, leads Socrates to the famous "second sailing" that consists of an investigation to the *logoi* of things, the latter being, as it is made clear in the dialogue, their final causes (thus, every form for Plato ought to be understood as a final cause). For the remaining pages of the dialogue, Plato singles out one particular form, that of beauty, which, by calling everything to itself, makes every thing that heeds its call—and everything to some degree is—beautiful. Joseph Pieper finds this Platonic insight captured by one of Goethe's sayings "Beauty is not so much a fulfillment as rather a promise." He, then, elaborates:

> In other words, by absorbing beauty with the right disposition, we experience, not gratification, satisfaction, and enjoyment but the arousal of an expectation; we are oriented toward something "not-yet-here." He who submits properly to the encounter with beauty will be given the sight and taste not of a fulfillment but of a promise—a promise that, in our bodily existence, can never be fulfilled.[48]

Dionysius (the Pseudo-Aeropagite) is situated in the middle of the distance between Plato and Kant. His beautiful is not anymore as impersonal as Plato's form, nor has it been yet depersonalized as Kant's *a priori* idea of purposiveness. For Dionysius the beautiful is a person, God himself:

> The beautiful [*kalon*] that is beyond all being is called beautiful [*kallos*] on account of its own beauty that it transmits to each and every thing and for being accountable for the harmony and brilliance of all as the light that shines to every thing its radiating rays and for calling [*kaloun*] everything to itself and gathering everything and in every respect, for which reason it has been called beautiful [*kallos*].[49]

Therefore, if the beautiful is recognized as beautiful it is because it renders itself visible (i.e., it "calls" to itself) and, by the same token, what is visible, what appears in appearance and by appearing "calls" to itself, is only the beautiful. Dionysius's passage distinguishes between these two (simultaneous) movements clearly: the beautiful radiates "like the light"—thus it renders everything visible, indeed it is the condition of visibility—but also recollects everything to itself, now strictly in its capacity as the "beautiful"—that is, as a call from the future.

It is this double movement of the beautiful/visible that Marion's phenomenology of saturation retrieves. What these phenomena are saturated with is the excess of the givenness of the phenomenon itself—it is an excess of intuition, a surplus of information we would say, that saturates them. This, however, does not mean that we have to look for saturated phenomena far, not, for sure, among the exotic, the extraordinary and perhaps the bizarre. Saturated phenomena are not a special group of phenomena but every phenomenon when seen without the protective glasses of regulatory concepts and preconceived intentionalities. Every phenomenon is inexhaustible—there is no viewing of a painting that is ever final, as there is no performance of a composition that is definitive; there is no event that can be transfixed into a single interpretation and, above all, there is no Other that would fit comfortably in one of my categories. We, now, understand that saturation is complemented by and, indeed, results to some kind of negation (negative theology). The task of the phenomenologist of the abundant givenness is similar to the theologian of the divine names: never-ending, or, as one could say after Gregory of Nyssa, *epectatic*. Everything gives always more than one can receive—it is this generosity of phenomenality that necessitates revision, repetition, interpretation and finally, what gives rise to philosophy itself, wonder. This fecundity of intuition surrounds every phenomenon as if it were a halo of excessive visibility—a metaphor often evoked by Husserl himself in his *Ideas*—a mandorla of light, that transforms phenomena—better yet, it renders them visible. For, phenomenologically speaking, in order to see what is seen one must also "see" what one cannot see, what remains unseen and as such shows the visible. There is no doubt that the theme of the abundance and irreducibility of donation as well as the chiastic intertwining of the visible and the invisible bear a strong affinity with a theological worldview. For ultimately the phenomenon of revelation conditions the revelation of the phenomena.

6. Perfection in imperfection

If, indeed, only the end (in the double sense of *telos* as finality and purposiveness) makes things perfect (*teleia*), then purpose keeps reminding us of such perfection amidst incompletion and imperfection. It is as if the human mind were indeed made in such a way as to understand only the perfect and the complete. For even if this is lacking in the present state of things, and it can only be lacking, then it feels compelled to supply it by itself. Memory and anticipation are both mediums of "idealization," that is, of bestowing perfection upon the thing remembered or expected that, once presented, the thing lacks. Hence the disenchantment that follows every realized expectation.

The perception of any physical thing, that is, any perception in space, "involves a certain *inadequacy*."[50] By this Husserl means that what I see is always necessarily partial, for I can never grasp what I see fully, from every single angle, in every possible way an object can show itself to me. This partiality, this imperfection endemic to perception itself, is a limitation necessitated by the limitations that are imposed on both me, as the perceiver, and the object of my perception, on account of our respective embodiments. However, the same imperfection perpetuates a series of inexhaustible possible perceptions "which can always be continued" and "which are never completed."[51]

Yet, even if the act of perception can never be completed, what is perceived cannot but be comprehended *as if* it were complete. For I never see the book that lies on my desk as the one-sided, two-dimensional patch of blue color on my visual field—*that* would amount to not seeing the book at all—but as an object in which all its characteristics, properties, and angles are somehow presented in a unified way, such that my gazing alone can never discover, and not of any shortcoming on its account, but precisely because at no given instant, at no given perception could any object be so presented. From where does such completeness come? The answer can only be: from the consciousness itself.[52]

"When I perceive simply, moving about in my environmental world, when I see houses, for example, I do not first see houses primarily and expressly in their individuation, in their distinctiveness [and thus, in their incompleteness]. Rather, I first see universally: this is a house."[53] What Heidegger alludes here is the eidetic intuition of "the essence of any empirically possible or impossible house" that is given together with the intuition of that particular house in front of me. As Marion explains: "I see *the* house, *as* house, before seeing (and in order to see) *a* house; or rather, the *as* of the house precedes a particular house and allows it to appear as such."[54] Indeed, intuition always gives more than one suspects—for it presents us the world at least twofold: in its particularity and in its universality; in its particularity *through* its universality. What I see, then, when I see this or that is never the particular thing in itself but the thing in its eidetic horizon without which that thing in itself would be, strictly speaking, invisible. What allows things to appear as the things they are is the surplus of such eidetic intuition, an intuition that regards their *eidos*. Yet such a regard is not a gaze upward as it was supposedly for the Platonic philosopher in search of his forms, but a gaze forward to the eschatological perfection of things. The *eidos* of a thing is the thing as given within a horizon of perfection and completion, that is, a state that a thing can have only at the end when completed and perfected. So we read in St. Augustine:

> True equality and similitude, true and primal unity, are not perceived by the eye of flesh or by any bodily sense, but are known by the mind. How is equality of any kind demanded in bodies, and how are we convinced that any equality that may be seen there is far different from perfect equality, unless the mind sees that which is perfect? If indeed that which is not made [*facta*] can be called perfect [*perfecta*].[55]

A little more needs to be said here by way of an explanation of temporality's role that, even though throughout implicit in the foregoing analysis, it has not yet been thematized as such. The very notion of perfection implies, on its most basic, etymological level, a terminus or a *telos* reached over a period of time and by means of such time (*per*-factum), and thus *finished*. Perfection is an end-of-time category. To say this does not necessary imply an absolute, "end of times" (in plural) scenario, although all teleologies draw their meaning from within such an eschatological perspective. It simply means that perfection as a finishing that has been now finished cannot be looked for at the beginning. No beginning *qua* beginning can be perfect. Perfection is inseparably connected to the notion of time and, more particularly, to time as time passed. To the phenomenological eye perfection is not presented by the things themselves—which, as we have seen, are always and necessarily given through inexhaustible albeit partial chiaroscuro of perception—but is supplemented by consciousness, a consciousness for which each and every of its cogitations are always equally necessarily presented in the flow of time:

> In itself *every mental process* [*Erlebnis*] is a flux of becoming, is what it is in a generation originaliter of an invariant essential type; it is a continuous flow of retentions and protentions mediated by a flowing phase of originarity itself in which there is consciousness of the living now of the mental process in contradistinction to its "before" and "after."[56]

The "continuous flow of retentions and protentions" that consciousness essentially is does not move in a simple linear fashion—as our naïve conceptions of time might have it—from past to present to future (or reversely), but they form a highly complex pattern where the three dimensions of time are interwoven perichoretically within each other, so as every now to contain a retention of the "having been" as much as a protention of the "about to be." In turn, each retention as well as each protention is pregnant with a similar tripartion of the now, the before and the after, and so on. One could give the example of comprehending the verse of a poem as one recites it: obviously one cannot utter all the syllables that make up any given verse—let alone all verses—of the poem at once, but during each one of them, as one sound is followed by another, the words spoken are retained in the words one now speaks, and the words yet unspoken are anticipated in the words one now speaks. The example is a favorite of St. Augustine who is using it in order to illustrate the passing of time in the famous discussion of time in the *Confessions*. Yet we find a more apt employment of the same metaphor in some other of his works. For example, in *De Vera Religione* he writes:

> A line of poetry is beautiful in its own way though no two syllables can be spoken at the same time. The second cannot be spoken till the first is finished. So in due order the end of the line is reached. When the last syllable is spoken the previous ones are not heard at the same time, and yet along with the preceding ones it makes the form and metrical arrangement *complete*.[57]

That St. Augustine is using some proto-phenomenological skills in his observations is confirmed by Maurice Merleau-Ponty who, centuries after Augustine, used this very metaphor about signification: we understand the beginning of a sentence from its end, Merleau-Ponty reminds us, as we understand movement in light of its teleological direction.[58]

7. The theological constitution of consciousness

Let us now bring to our discussion a third interlocutor, Søren Kierkegaard, who in the *Concept of Anxiety* offers the following gloss on Romans 8:19: "inasmuch as one can speak of an eager longing, it follows as a matter of course that the creation is in a state of imperfection."[59] As we have already seen in the first section, this *longing* could not be understood except in terms of direction, movement, and desire that stretches out to its ultimate good? What is this "ultimate good"? Let's call it, after Kierkegaard, "perfection." Thus movement, in this case disguised as desire, becomes the indication of the *expectation* of a future perfection (perfection can only be futural) *and*, at the same time the means by which this perfection is to be carried out.

Here two elements come together: *anticipation* (desire as movement, expectation, the future) and *pleasure* (desire as perfection: "all pleasures have within themselves some feeling of perfection" wrote Leibniz[60])—the anticipation of pleasure, as well as its reverse side, (namely, the pleasure of anticipation, the foreplay), unite the temporal and the bodily insofar as, in its most immediate level, this anticipation of pleasure is the expectation for the Other's body: both sexual and Eucharistic at once. At their crisscrossing, the pleasure of anticipation and the anticipation of pleasure dislocate the present as the *privileged* vantage point in the metaphysics of time (from Aristotle to Hegel and beyond).[61]

How are we to understand this ability of the mind to see perfection when perfection is lacking? It is precisely at this point that we need to turn to a phenomenological inquiry of the teleological. It would seem that the first (that is, the most fundamental and the most readily available) intuition of eschatology is that of awaiting or expecting ("the hoped for" or even the "unhoped for" as in the work of Jean-Louis Chrétien). But what would such an intuition have been without the idea of purpose, that is, of fulfillment of one's anticipation, even if we were not to know what or whom we are waiting for? More fundamental then than waiting is this waiting-for, that is, the structure of a purpose. Whence can we phenomenologically derive such a structure? First of all, from the very character of intending. Intentionality, even prior to intending this or that, always intends a purpose; in fact, it is purposive.[62] In every fulfillment, in every filled intention, one can observe the structure of the teleological. Kant spoke of pleasure precisely on these terms[63] and we believe that it is the joy of the kingdom to come that is foreshadowed in the feeling of satisfaction that every filled anticipation yields.[64] "All pleasures," after all "have within themselves some feeling of perfection"[65] and, St. Thomas Aquinas continues, "in desiring its own perfection everything is desiring God himself."[66] The very passage from an empty intention to a filled one, that is, the passage from absence to presence, is such a teleological indication for in all these common structures of anticipation the absolute anticipation, i.e., the anticipation of the absolute, is reflected.[67]

Teleology is, of course, one of the oldest "proofs" of the existence of God in the books: from St. Thomas Aquinas's fifth way to Leibniz's "principle of the best." Nevertheless, it was believed to have been entirely discredited as an argument when it was shown that purposiveness cannot objectively be found "out there"—that is, independently of the human mind. It must, therefore, be *only* in the mind and a thing of the mind. Nothing would delight us more than this conclusion. Its supporter might have thought that achieved a decisive blow against teleology, without realizing that he had furnished it instead with its strongest defense. For to say that teleology is a property of the mind is to elevate, as we have tried to show, to a universal structure of human consciousness. Consciousness first projects perfection in the world and only then, on the basis of such projection, discovers a teleological perfection in the things themselves. So far consciousness had failed to recognize purposiveness as its own essential characteristic and had, therefore, mistook it as a property of a world assumed to be external to itself. Let this mistake be corrected by the Copernican Revolution that phenomenology brings about and let teleology, so understood, become the first indication of what could be called the *theological constitution of consciousness.*

Part Two

The Scandal of the Good

The First Garden: *Being at the Beginning*

3

In the Beginning

I. The unconditional and the uncreated

"And this narration itself constitutes the problem: who speaks, about what precisely, and to whom?" writes Marion. And he continues, "the most inadequate solution consists in falling back on something apparently obvious, by assigning the *Confessions* the status of an autobiography, without worrying anymore about the *autos*, the *self*, of the question."[1] Our task for now is precisely such "worrying": and let us begin by thinking the difference between a confession and what the *Confessions* are normally taken to be, an autobiography, by following the indication of Marion's three questions: "who speaks, about what precisely, and to whom?" If, then, the discussion is about autobiography understood by what the term itself gives to understanding, we should begin by taking notice of the "*auto*" of autobiography as pointing to a *self*-narrated life (*bios*): a self-narrated self as opposed to biography in which one lets the story of his life, of his self, be told by another. The prerogative of autobiography is simply that the self-narrated self sets down its own rules and regulations first of narration as well as of living. The *auto* governs the *bios* (life) as much as the *–graphy* (written account). I reserve the right to tell the story of my life and to tell it as I like, that is, as I understand it, as I give meaning to it, and, as autobiography claims, I give its meaning. Thus, the *auto* of autobiography does not merely tell the story of itself but gives also its "law"—that is, it decides and determines what will be said and how it will be said, what will be read and, as much as possible, how it should be understood. In this ambition, implicit in the autobiographical effort, the self reveals itself as a self that sets for itself (if not also for others) its own rules, its own *nomos*, thus *autonomous*. Following one by one the layers of auto-ism (if not autism) that autobiography has opened for us, we discovered that such an autonomous self implies an independence from anyone and anything, a self that is not indebted to anything other than itself, in as much as it is not under any principle than its own or that of its autonomy. Principle here is meant in the double sense of *principium*, that is, of its *raison d'être* as well as of the beginning, of its *archē*: a self given to itself, self-given, that is, *autarchic*. Here we have then a triad of closely interrelated concepts: *autobiography*, *autonomy*, *autarchy*.

Does this picture, however, of a self-narrated, self-reliable, self-regulated self square with the openness to another that the act of narration presupposes? Every narration is a story, and a story is something told to someone about something.[2] Story-telling is other-dependent. Was it not this the point illustrated, for example, by the structure of the *Symposium*, encased as it is in the double indirect speech of layers upon layers of narration? As Marion says about confession, "thinking amounts to thinking in the more originary mode of confession of . . ., toward . . ., for . . . —an other instance besides myself, besides the *self*."[3] What can possibly create the need or even the motive for the autarchic and autonomous self for its own autobiography? Is there a motive of telling who I am in the closed universe of solipsism? That is, from the *inside*, an inside that lacks every exteriority, what possibility is there for story-telling, even for the story-telling about oneself, the self-telling of the self's story?

Here already we hear the problematic of a different question, much larger than the seemingly technical question of the possibility to a self-narrated self: it is the difficulty of justifying creation from the point of view of a self-sufficient God, that is, of God without relation and outside of communion. The creation of the world stands here in parallelism to the narration of the self. Remember the beginning of Psalm (18/19) "The heavens narrate [διηγοῦνται/enarrant] the glory of God/and the firmament shows [ἀναγγέλλει/adnuntiat] His handiwork [ποίησιν/opera]"—the "poem" that the firmament announces and the glory that the heavens narrate, that is, in the hendiadys of the psalm, the "glorious poem," is nothing other than themselves, their own making. Augustine would later dramatize this Psalmist verse when the heavens respond to his intellectual queries precisely by referring him to their creator: *ipse fecit nos* (X, 6, 9 [pp. 242–3]). Similarly, in book XI, their temporality proclaims that they were made and that they "did not exist before [they] came to be, as though to bring [themselves] into being" (XI, 4, 6 [p. 288]). And in the allegorical interpretation of book XIII, the firmament is an image of the Holy Scripture itself, a scroll stretching from one end of the horizon to the other, a canopy of text (XIII, 15, 16 [pp. 353–4]). How could or rather, why a strictly monotheistic God create at all?

We return to the question of autobiography in order to declare its impossibility. Not only "the *Confessions* should not, above all, be read as an *auto*biography"[4] but, autobiography as such is impossible. Any attempt to it should be only understood as the illusion that a self-enchanted self suffers in thinking itself the source and cause of itself: of its self-narration first, but also of its life.

Autobiography is *twice* impossible: *formally*, insofar as it violates in principle the syntagmatic orders of narrative by remaining inexplicable with regards to its movement, its motives, its direction, but more emphatically *objectively*. In other words, even if we were to grant to the autonomous self the possibility of self-narrating itself to itself, the question remains: *from where would it have to begin*? I would like to tell you the story of myself, but I do not know from where to begin. Does one begin always from the beginning? But the autarchic self—the only one that could lay a claim to autobiography—has claimed itself as the beginning of itself, that is, as having no beginning. An autobiography that begins is travesty. An autobiography of a being that has begun, a being with a beginning, is a contradiction.[5] Even in the case of the creation of the world (a problematic that doubles that of the narration of the *Confessions* and which culminates in book XI), the demand for a beginning is preserved in all its force. The whole of creation is a being that is above all temporal, that is, in time, and as such, it must have a beginning. "In the *beginning* God created heaven and earth" (Gen. 1:1). Yet, how could the eternal God create that which, by necessity, that is, by its nature, would have to be in time, and thus begun? The answer given in book XI, to which here we can only allude, is precisely that at which our question above had hinted: namely, that of a God who creates through his co-eternal and consubstantial Word, Who is also the Beginning (John 8:25), yet as an *eternal* beginning, He is himself a beginning without beginning. In the Holy Trinity, both the demand of a creation that began, and of a God who preserves his eternity are satisfied. Nevertheless, there is no recourse to a community that the autarchic self, *qua* autarchic, can have, and therefore, without beginning is also without the Word/word, without the creative speech, without the creative power: mute. Autobiography is impossible.

Today the celebrities who write their own autobiography wish to tell us their life-stories as *they* know them, for they assume that no one knows them better than themselves. Perhaps, this would be so, provided that one is willing to restrict one's life only to the anecdotal, comprised by a series of dark secrets. Aristotle, however, had stipulated that one is known better by another and that it was necessary for one's self-knowledge to have friends, for through the friend, as in a mirror, one could come to see oneself.[6] Leaving Aristotle aside, it would be hard to maintain the presumption that I know better myself, when I had to be told who I am by others. Since, in spite the illusion of autarchy, I was not my own beginning, nor was I at my beginning, but rather, in many ways, I come after myself, I had to be told who I was by my parents and relatives. The three calls of givenness, as they were reviewed in the second chapter of our prolegomena, have already anticipated that the first story of myself, like my self, was given.

By raising the inevitable problem that the beginning posits for autobiography, as much as an autobiography properly understood must be a work without beginning and unable to begin, we have already alluded to the crucial issue of *time*. Following on the foregoing argument, autobiography belongs to the horizon of *synchronicity*—that is, it knows of no time difference, which means it knows of no time at all. On the other hand, *confession* belongs fundamentally to the mode of *diachronicity*—it is possible *only* through the opening of time, enabled by the difference between a now and a then, between a before and an after.

Yet, the *Confessions* begin and appropriately not *sub propria voce sui* but through the voice of another, that of the psalmist. Book One speaks of the origins of man, not only of *this* man who recalls his childhood in Thagaste, but of man in general, of mankind as such, and therefore, of the origins of sin. Already in the opening lines the theme of death and sin—and of their connection that is traced back to the primordial Garden which St. Augustine's narrative so skillfully re-enacts in the Garden of the stolen fruit—that is, the theme of death as the result of sin is announced: "we who carry our mortality about with us, carry the evidence of our sin and with it the proof that you thwart the proud" (I.1.1, p. 39). *Pride*, namely the refusal to recognize one's origin, or better yet the refusal to recognize *that* one has an origin,[7] becomes the *beginning* of sin. So here sin begins with the denial of the beginning. Thus the *Wisdom of Sirach* affirms that "pride is the beginning [the *archē*] of all sin" (10:13).[8] Let it be noted that pride is, incidentally, autobiography's motivating thrust which, it should be now made clear, far from being synonymous with confession, is its very opposite.

Confession: all three meanings of the term "confession" form a conceptual cluster: a) confession of sins—it is only this that can give some justification to the anecdotal aspect of the autobiographical; b) confession as profession of faith (as we say for many saints that there were confessors); and c) confession as praise. In realizing one's sinfulness (a) one feels gratitude for God's salvific plan and feels the need to exalt God (c) for His goodness (ἐξομολογήσθε τῷ Κυρίῳ ὅτι ἀγαθός/confitemini Domino quoniam bonus, Ps. 117/8), thereby one's belief in God's love becomes firm (b). Therefore, there is a logical progression that leads from the one to the other or a relationship that binds these three meanings together.

"O give thanks [ἐξομολογήσθε/confitemini] unto the Lord; for he is good: because his mercy endures for ever." The opening verse of Ps. 117/8 gives us the conditions, if you would like, of confession as doxology. You are brought to praise when you realize that the Lord "is good." From what perspective can one know that the Lord "is good" and, more significantly, what does it mean to say that "his mercy endures for ever"? We cannot take the knowledge of God's goodness as self-evident, that is to say, without presuppositions. To ask a similar question: what motivated St. Augustine in writing the *Confessions* and what did he hope to achieve by doing so? He tells us in his *Retractiones*, to "arouse the human mind and affections toward [God]." How so? By showing us that God's "mercy endures for ever," in spite of his sins, and thus prompting us, his readers, to "give thanks to the Lord, for he is good." Again, confession progresses in these three moments:

a) Reflection of one's sins that leads to the realization that I am sinful (we shall see in a moment what is the fundamental structure of this realization)—*metanoia*.
b) Together with my sinfulness, I become aware that, even though sinful, I am not completely abandoned by God, that His mercy *endures* my sinfulness. This, in turn, moves me to:
c) Realization that the "Lord is good," which is a confession of faith for it makes a statement about what God is (theology) as much as doxological statement (ultimately, all theology is doxology).

In each of these moments I catch sight of myself, my life, as manifested to me over a period of time. That is to say, I see myself as *enduring*, like God's mercy, or because of it. I see myself as "having been here before" or "having done this before." To become mindful of one's sin, indeed, the very concept of sin, depends upon repetition. We feel the strength of sin more clearly in its compulsion. The threefold aspect of confession—realization of sinfulness, realization of God's mercy, gratitude—opens ourselves to time. It is precisely such a temporal element that the *meta* of *metanoia* indicates so succinctly. When the mind is affected by the *meta* of time then we can speak, in the Latin idiom, of a "transformation" as we often hear about such persons that they were a transformed or changed person.

To begin with a prayer—as the *Confessions* do by combining a number of verses from the Psalms—is to circumvent the problem of the beginning, for a prayer is never mine. It predates me, as the Psalms of David predate Augustine's *Confessions*, and, therefore, when the latter begin by a citation of the former, the book hails back to a time before its beginning, it places, or rather dis-places itself to a beginning that has always already begun, a beginning immemorial (the significance of this will become more evident later on in the course of remembering, or more correctly, failing to remember the beginning of his infancy, and even that other beginning that possibly predated it). In this sense, every prayer is like Charles Ryder's prayer, "an ancient, newly learned form of words."[9] Every prayer is both ancient and new, like Augustine's God ("Late have I loved you, Beauty so ancient and so new" (X.27.38, 262)). And in so far the words of the Scripture, in this work in particular, are the words of God, the words that the Word speaks, as the double imperative "*tolle, lege*" would exemplify later, then this initial address *to* God is made possible only when one speaks God's words: when one prays. Most of the prayers a Christian would say even today are citations—citations of God. Think, for example, of "our Father." Every time we recite this prayer one cites the words of the Word—for prayer too, like praise ("you stir us [to] praising" (I.1.1)), like faith ("this faith which is your gift to me," 1.1.1, p. 39) are given to us.[10] In whoever prays it is God who prays with him.

Yet, what does it mean that the words of my prayer predate me, if not that they speak me before I speak them? The language of prayer, which is also the language of the community within which I am born or reborn, comes before me, and makes the effort to my own beginning already vain. For if I am to begin *speaking*—to return to the impossibility of autobiography—I will have to do so in someone else's words. No one invents one's own language and such a language, as strictly private, would have also been strictly *idiotic*. Thus, the very words that one may use to affirm his autonomy fail him and betray his dependence in what is always prior and anterior than the self: language. God has spoken first and humanity's word, whether one realizes it or not, comes always as a response to that first word. Humankind's language is responsorial.

The beginning signals the impossibility of a beginning—especially for us who have already begun, *in media res*. The steps in the opening phrase: mortality a result of sin, sin a result of pride (rather pride is the source of all sin, the *criterion* of sinfulness, what makes sin "sinful"). Pride solidifies sin by almost precluding the possibility of salvation, for if man in his pride fails to recognize sin as his and as sin, that is, as a sin, what salvation can there be? I am suspecting that the "proof that you thwart the proud/*testimonium, quia superbis resistis*" of the opening paragraph might hide a reference to the Manichean presumption for they have thought of themselves as equal to God and by positing an evil principle they had exculpated themselves, precluding, thus, the possibility of confession, or rather, of the humility that can lead one to confession. Indeed, the second half of the verse cited here (Prov. 3:34) remains, for the time being, suppressed. Even so, it is brought all the more to our minds: "but to the humble he gives grace/*humilibus autem dat gratiam*." Grace is left unspoken, for it will be shown, indeed, it will shine forth, in the course of the *Confessions*. Manichean *superbia* was doubled by the *curiositas* that characterized their fanciful genealogies (cf., Paul's warning to Timothy against those people who "teach false doctrines ... or devote themselves to myths and endless genealogies" 1 Tim. 1:4)[11] indicative of a rationalism that took pride in its knowledge (*gnosis*), making knowledge, instead of faith (*pistis*), its point of departure. Like Zacchaeus, in the Gospel story (Lk. 19), the Manicheans had climbed the fig-tree of the pseudonymous knowledge of the good and evil motivated solely by intellectual curiosity (*concupiscentia oculorum*); yet, Jesus command to Zacchaeus was to "come down" (Lk. 19:5). For finding God and entering into His kingdom requires that we become "little" again, like Zacchaeus, who was "little of stature" (Lk. 19:3). It is this, and not some innate innocence, that Augustine saw in Christ's declaration with regards to the little children that "of such is the kingdom of heaven" (Mt. 19:4): "It was only the small stature of a child that you mentioned with approval as a symbol of humility, O Lord our king, when you declared that *of such is the kingdom of heaven*" (I.19.30, p. 60). The prooemium of Book I comes to a close with another echo to Luke's account of Zacchaeus encounter with the Lord. As Jesus said "Zacchaeus, make haste, and come down; for today I must abide at thy house" (Lk. 19:5), so Augustine asks of Him: "The house of my soul is too small for you to enter: make it more spacious by your coming. It lies in ruins: rebuild it. Some things are to be found there which will offend your gaze; I confess this to be so and know it well. But who will clean my house?" (I.5.6, p. 42).

Book One of the *Confessions* speaks of a self already given (hence the impossibility of self-beginning). It speaks of the origins of man and of original sin. The opening of the book does not leave any doubts that for St. Augustine what we have called after him "the original sin" is identified with pride: *homo circumferens mortalitatem suam, circumferens testimonium peccati sui*. Death is the testimony of humankind's sin but, at the same time, *testimonium, quia superbis resistis*—that "you thwart the proud." Why, then, is pride the origin of all sin, the original sin? Precisely because it refuses to recognize one's origin: first it fails to recognize that one has an origin, and that therefore we are not anarchic, and secondly, it fails to recognize who this origin is. Book One tells us that the self is given. At the coda at the end of this book, listing the gifts that he has received, Augustine declares them to be just that: "all these things are gifts from my God" (I.20.31, p. 60). And this is, naturally, a cause for thanksgiving: "I give thanks to you for your gifts" (I.20.31, p. 60). More important, however, are the concluding words: "because this too is your gift to me—that I exist" (*quia et ut sim tu dedisti mihi*, I.20.31, p. 61). Being itself, my being, has been given to me.

The fundamental nature of humankind as essentially given—a "being given" to invoke the title of one of Marion's books—can perhaps be best explained by its creaturely character, that is, as being created and belonging to creation. A few words, then, on the phenomenology of creation are in order.

II. On the phenomenology of creation

Lacoste's analysis in *Experience and the Absolute* rests on the phenomenological principle that both, the world and consciousness, are given simultaneously and the one emerges only by a "pro-vocation" of the other; by a reciprocal donation that remains, in terms of its origin, inexplicable. At its root lies precisely the enigma of origin as the possibility of what is possible—an enigma which, as we shall see, can be understood only from the end. The enigma of the beginning is nothing else than the question of time, and time—the passing of time, time as passing—acquires meaning only once it has passed, once it has been completed, made [*factum*] through [*per*], and, thus, made *perfect*, for the beginning and what begins. In short, the *penultimate*, draws its existence and its significance from the perfection of the *ultimate*.

More about all of this, however, later. For now, let it suffice to note that a phenomenology of creation cannot but be posited at the same time with, and as the corollary of, a phenomenology of the self. It is not accidental that in Marion's recent discussion of St. Augustine, the last section of chapter 5 entitled "the event of creation" finds its continuity in the next chapter that bears the title "The Creation of the Self." What does the self have in common with creation? On the one hand, we have "the impossibility of the individual event, which should alone decide all that I am, my ipseity and my identity—the event of my 'conversion,' as a decision for God."[12] On the other hand, there is "the actuality of the cosmic event, the emergence of the heavens and the earth, an event always already decided, but in such a way that it imposes itself as a fact irremediably accomplished, but also definitively incomprehensible, since it precedes me by a radical facticity, a saturated phenomenon."[13] Despite their mirroring juxtaposition, or rather because of it, "what these two events have in common" is precisely "creation"—"the twofold event of creation" [that is, the creation of the world and the creation of the self].[14] Said otherwise, what is common in these two events "is no longer even named an event but an advent—the advent of time itself."[15]

The reason for this necessity has been already indicated above: the meaning of creation lies at its beginning, that is, at the fact that it had a beginning: "In the beginning God created heaven and earth" (Genesis 1:1). Similarly, the question of the self—and that includes the self as a question for itself which is fundamentally what the self is for itself[16]—lies also in its beginning, a fact that becomes more pronounced as long as humankind tries to become its own beginning. It is far from accidental that when Bonhoeffer discusses the beginning of creation in his commentary on the first verse of *Genesis*, he passes naturally, and without feeling the need to explain this transition, to a discussion of man's beginning and of man's relation to the beginning.[17]

What is the world? Usually, we know what we mean by this term without the need for further explanation or definition. The world, one could say, is everything that is "out there" and "around" us, people and things that we encounter in our everyday lives. Yet we understand that the world is not any of these things separately, nor is it all these things together. Simply put, the world is neither a thing nor a collection of things. Phenomenologically speaking, the world is not a phenomenon but the horizon within which phenomena appear.[18]

The "natural" attitude of humankind toward the surrounding world is to find itself in it. The first concept of the world is that man is *in* it.[19] This understanding seems to reveal a certain image of the world as a container, as an extension that stretches to immeasurable lengths. Here we already have the modern understanding of the world in terms of *space*. Against such a view of the world, we could contrast our everyday being-in-the-world: the totality of lived experiences that make the world more than a space, that make it our *place*. Although that first understanding of the world as space seemed natural, in fact we never live by it; it is an understanding that can be reached only once we have detached ourselves from our experience of the world as place. What we have called here a "natural" attitude is really *un*-natural,[20] for it can be formulated only through an artificial stance of disentanglement from our relations with the world. But the world is nothing other than precisely this entanglement—a nexus of references. We are always involved in the world in such a way that there is no world without, or apart from, these involvements. It is humanity (our affairs) that makes the world worldly, that is, more than a container in which humankind finds itself as a fish finds itself in an aquarium.[21]

Space and place: these terms are not theological, although the distinction that differentiates them is. That distinction is introduced by Christian thought as between created and uncreated orders. It is interesting that Heidegger recalls this distinction precisely in his discussion of modernity's inception with Cartesian dualism (*Being and Time*, Division One, III, B), yet he dismisses it for the sake of the ontological difference (another kind of dualism perhaps?) according to which "what is created and the creator alike [are] *beings*."[22] Please note that this distinction between created and uncreated orders is absent from classical thought, and it has been eclipsed in modern times—although it is upon this distinction that the inception of modernity rests, with John Philoponus's critique on Aristotelian cosmology.[23] Thus modernity never thinks or takes into account the difference upon which it is itself constituted. Modernity, in a way, becomes forgetful of its own origins. The image of a spatially understood world confuses the distinction between *ens creatum* and *ens increatum* by substituting it with a differentiation between the human and the natural. Aristotle's philosophy operates precisely on such terms by distinguishing between what is from man (that is, *technē*) and what is independent of human craftsmanship (that is, *physei*).[24] The very concept of *physis* makes sense only within such a scheme that distinguishes the natural from the artificial. However, such a perspective fails to recognize that physis *is not* ktisis. The concept of creation (*ktisis*)[25] places humanity together with the rest of creation indistinguishably—creation is humankind and world thought together; thus, the world becomes humanity's place. The following anecdote from one of Lacan's seminars provides us with further corroboration:

> I remember that one evening when I was dining at the home of a descendant of one of those royal bankers who welcomed Heinrich Heine to Paris just over a century ago, I astonished him by telling him—he remains astonished up to this day, and is still clearly not ready to get over it—that modern science, the kind that was born with Galileo, could only have developed out of biblical or Judaic ideology, and not out of ancient philosophy or the Aristotelian tradition.[26]

Contrary to creation, nature as that which keeps flowing out of a primordial source and that primordial source itself that remains "hidden"[27] is itself its own origin or, rather, the lack of origin (*Abgrund*). Several ideas find their birthplace in this image of the aboriginal flowing of nature: for example, nature's anarchy and thus eternity, nature's emanation, and nature's divinity. Thus, ancient polytheism as it was structured around this cluster of concepts associated with nature becomes, over time, translated to the modern *a priori* atheism of being-in-the-world. What these ancient ideas held in common was an understanding of nature as *necessary*. For classical thought, *physis* exists necessarily and its necessary character scorns humankind's contingency. On the other hand, creation, precisely in its beginning, underscores *freedom*. That freedom is first and foremost exemplified in creation's gratuitousness. The *ex nihilo* of the creation—the world as created as much as the creative act by which "[t]here is no possible question that could go back behind this God who created in the beginning . . . No question can go back behind the creating god, because one cannot go back behind the beginning."[28] That is because, according to Bonhoeffer, the beginning "is completely free."

> In other words the Creator—in freedom!—creates the creature. The connection between them is conditioned by nothing except freedom, which means that it is unconditioned.[29]

In this assertion, Bonhoeffer follows the Patristic tradition of the Church for which the creation of the world was always unconditioned and ungrounded, precisely in order to safeguard the freedom of God's creative act. Every compromise on this score, usually erring on the side of Platonism (cf., Origen, Schelling) meant

> . . . that God was a creator by necessity and not freely. Without creating the world God would remain unfulfilled, he would not be God. The notion of God and the notion of creation thus overlap, and paganism makes its appearance disguised under the form of Christian doctrine.[30]

It is precisely this kind of disguise that one encounters, for example, in Schelling's positing of an eternal yearning or craving (*Sehnsucht*) in God that necessitates God's self-manifestation in the "creation" of the world.[31] Going further than the ontological necessity, Schelling does not hesitate to speculate the world's coming-to-be as "morally necessary" (*eine sittlich-notwendige Tat*).[32]

It would be useful to remember here Kierkegaard's argument that no becoming can be necessary, and what has been does not become so by having been (i.e., as *fact*) but, precisely in having been, it demonstrates that it was *not* necessary. So with the creation: if it came-to-be, its very be-coming—what we signified above with the *beginning*—implies the lack of necessity and, in contradistinction, a passing from possibility to actuality that, for Kierkegaard, "takes place in freedom."[33]

> Necessity stands all by itself. Nothing whatever comes into existence by way of necessity, no more than necessity comes into existence or anything in coming into existence becomes the necessary. Nothing whatever exists because it is necessary . . . All coming into existence occurs in freedom, not by way of necessity. Nothing coming into existence comes into existence by way of a ground, but everything by way of cause. Every cause ends in a freely acting cause.[34]

Phenomenologically, then, we never encounter the world as nature—in that metaphysical definition of which this term has been employed here—but always as creation: that is, we encounter a world of becoming and of change. And even though one could not possibly have access to that beginning of which the first verse of *Genesis* speaks, nevertheless that beginning reverberates in every one of our engagements with the world insofar as in all of them our perception and our imagination meets the world against the background of possibility, a possibility that is always and ultimately a possibility of change and for change, the same possibility that makes the encounter with things in the world possible and which phenomenology simply calls *phenomenality*. Finally, it is the same possibility that, in Heidegger's famous dictum, stands higher than actuality[35] and which we earlier called *contingency*[36]—that is, *the beginning*.

The difference between these two understanding of the world, namely the classical notion of nature and the phenomenological description of being-in-the-world is decided above all by the place time occupies in each of them. While virtually absent in the former, time plays a central role for phenomenology. This importance is, in turn, attested by the systematic analysis that time received by the father of the phenomenological method, Edmund Husserl, in his study *On the Phenomenology of the Consciousness of Internal Time*. For the beginning of creation is the beginning of time.

According to the principle of reciprocity between world and self with which we opened this chapter, we could now move from these initial remarks on the phenomenology of creation to discovering their correspondence, so to speak, in the life of consciousness itself. The first such correspondence cannot be anything else but the evidence, present with and in every presence, of the realization of my self as a being that has already begun, that is, as a being unable to be present at its own beginning. Every attempt to do so has to be mediated by memory, that is, it has to be given in that special mode of remembering and of retrieving that of which I have no originary experience.

For sure, I can initiate new acts of consciousness but I cannot initiate consciousness itself. A consciousness that begins is a paradox. Yet, this does not imply consciousness's eternity but rather the very opposite: a heteronomic beginning that, once began, it cannot be surpassed, transcended, or passed beyond by consciousness. The beginning conditions consciousness and as such it constitutes for consciousness an unsurpassable limit. Further evidence can be gathered from what Husserl calls the "narrowness" of consciousness: "To narrow consciousness only a region of being appears, primarily because it has an intuition of something. Consciousness is narrow, also, because it can perceive remember, or imagine, but it cannot perceive, remember, and imagine simultaneously."[37]

We can hear "of the beginning only in the middle."[38] This phrase gives rise to the expression "man in the middle" which Bonhoeffer often employs to signify man's absolute latency with respect to his beginning (origin). One is not only always late for his appointment with God ("Late have I loved you, Beauty so ancient and so new"[39]), but also he is late with himself (his self). The two are not, of course, unrelated. We shall see later on that the Greek language captures well the existential dimension of this lateness, for one could say about humankind—as indeed the Gospel of Luke says about the prodigal son—that ὑστερεῖται, denoting at once both, the existential condition being-in-want and being-late. Man is behind himself and thus finds himself always in need. The interesting fact about us is that we do not quite coincide with ourselves and *desire* might just be the name of that disjuncture. As Stanislas Breton observes: "We do not coincide with ourselves. We exist before we are conscious of our existence; and this means that our reflective consciousness is always to some extent *out of joint* with the existential conditions that fostered it."[40] Or, after Merleau-Ponty, "Temporality both prevents my absolute coincidence with myself and yet is the means of establishing my own reality as myself."[41]

Phenomenologically, then, the world and its interpretation are presupposed for the individual consciousness. I find myself in *media res*, that is, already in-the-world, but not (at least not with the same sense of evidence) before-God. God is neither the world nor is he in-the-world. God is not to be found in the faithful execution of steps. My inheritance with the world by means of my body and its place remains *a priori* atheistic.[42] Yet, as a being who has already begun and for whom every attempt to begin (begin oneself), to be my beginning is doomed as impossible and comical, my origin, which is neither me nor my world (itself originated) remains hidden from me, perhaps by this very world in which I dwell, by my dwelling in the world; yet, it is at the same time revealed as hidden by the passing of everything I encounter in the world and by my mind's opening to time, to the passing of time, to time as passing. "The inadequacies of nature and history are God's cloak."[43]

III. The distraction and attraction of time

One of phenomenology's preferential examples of time has been music. The example of music was of course used by Augustine when he came to ponder the enigma to time, but also, more recently, it became paradigmatic for the analyses of temporality attempted by Husserl, Sartre, Levinas, and Lacoste. There are good reasons that suggest music as an ideal experience of time: indeed we know, on Kierkegaard's authority, that "all other media have space as their element. Only music ... occurs in time."[44] In respecting, thus, this long tradition in the history of philosophy, I would like to preface my comments on time with a brief analysis of well-known piece of music. What follows, then, is a brief theological reading of music.

I am guided in my analysis by a seemingly irrelevant question: that of the classification of Beethoven's composition that is known as his Ninth "Symphony." Let us assume a complete unfamiliarity with this work—bracketing out, so to speak, the countless times of listening to it and the very fact that it is now impossible to approach this work without knowing already too much about it: for example, that it is a "symphony" and the last one in the composition of symphonies by the great master. Let us try to re-create an introduction to it as if it were a first time. Imagine that it is Friday, May 7, 1824[45] and that we have taken our seat at the concert hall and already before the program has begun the work has opened to us and for us the question of its classification. We do not know any more what we should expect, what *kind* of music we are about to hear: for the members of the chorus have taken their position in front of the orchestra.[46] In the universe of well-defined musical genres one knows that a symphony is an orchestral piece of music—yet, here we have a chorus and four vocal soloists sitting patiently. What, at the beginning, we had assumed to be a symphony to be played in four movements by an orchestra, has now become (according to a little too descriptive and a little deceptive title of the program) "a symphony with chorus." That is, *more* than a symphony: but, then, could it still be called a symphony? It is as if one wanted to be clever and called an opera "a symphony with performed songs."

The music has now begun. Once more it surprises us, for what lies at its beginning and as its beginning is the lack of a beginning. We hear instead a sound that, in so many ways, is reminiscent of the sound an orchestra makes *before* it has begun. The beginning sounds like an orchestra tuning, getting ready to begin. Has it already begun? The distinction between the before and after of the beginning (and no distinction can be sharper than this) is here blurred so much as to leave one suspended for a while in the uncertainty of knowing not whether the work has begun. A beginning that begins after it has already begun: in the amorphous sonority of its first bars that give the impression of pure sound—of sound without articulation, of sound without form.

So far we have a symphony that is not a symphony (or, at least, not *only* a symphony) and a beginning that lacks all decisiveness of a beginning.

As the work progresses, we are able to recognize a more traditional arrangement of its movements, until we reach the final movement, when, once again, we are confronted with some notable perplexities. For at the beginning of the "symphony's" end we hear again the beginning—and more than that: all three preceding movements are rehearsed one by one at what is supposed to be itself a beginning of the final movement. In this rehearsal within the performance of the work, the orchestra seems to engage in a dialogue with itself, suggesting and rejecting a retrospective look in the composition's past. No: if this movement is to be the finale of this enigmatic work, then, it cannot return to its beginnings, rather its end must constitute something new, something that the three preceding movements, in all their originality and variety, could not have anticipated. Thus we imagine the cellos and basses speak as they dismiss one by one the orchestra's suggested themes, until the famous melody of the *Ode to Joy* is foreshadowed in the tenderness of a pianissimo. When later in the fourth movement the cellos' voice are translated in human language, then they indeed speak to us thus: *O Freunde, nicht diese Töne!*

From the nebulous beginning of formless sound and the explosive articulation of the first movement to the introduction of language in the finale, we can begin to appreciate that what Beethoven offers us in this composition is indeed "a work of cosmic proportion"[47]—for the scope of the work is no other, in the opinion of this listener, than that of creation from its beginning to its culmination which, as the emphasis of the chorus reminds us in the climax of its song, is "before God!" [*vor Gott!*]. Indeed, what we have here is an unconventional *oratorio*, cleverly disguised under the guises of a symphony: a very ambitious oratorio, as one would only expect from Beethoven, whose narration is not any particular event of sacred history, but the history of creation itself.

Nevertheless, Beethoven is not alone in undertaking such an ambitious project. St. Augustine, when he gives us a hint of the scope of his masterpiece, *The Confessions*, seems to have in mind a similarly cosmic perspective: "from the beginning when you made heaven and earth [the subject matter of Books XI and XII of the *Confessions*] to that everlasting reign when we shall be with you in your holy city [the topic of Book XIII]."[48] Curiously, Book XI is a book which, like the fourth movement of Beethoven's Ninth "Symphony," begins by rehearsing previous beginnings, in particular, the beginning of Book I ("Great is the Lord, and exceeding worthy of praise")—itself a non-beginning as it is a quotation of a number of psalms, thus echoing what has already begun—as well as the beginning of Book II ("that it is out of love for loving you that I do this"). The similarities and the parallels between these two great works of art do not stop here, yet for our purposes, it would be better to leave the analysis of Beethoven's music behind us at this point and move on to what Augustine's work has already suggested: living in the time between two ends. Reversing our methodology of reading music theologically, let us instead endeavor to read theology, the *Confessions* in particular, musically.

The famous problematic of the disjunction in the narrative of the *Confessions*—the apparent disparity between the two halves of the book (namely, in at least one count, between Books I–IX and Books XI–XIII) could be resolve with the same understanding that would refuse to see in the final movement of Beethoven's Ninth "Symphony" a redundant doubling in the presentation of the material, first by the orchestra alone, and then by the introduction of the chorus. Assuming that we were given to read the *Confessions* one book at the time, we should not expect another installment after the dramatic conversion of Book VIII or, at the very most, after the resolution reached by Augustine's baptism in Book IX.[49] In terms of dramaturgy, there is nothing more to be added or be told after these books. In terms of interpretation, the same point is asserted even more emphatically: for in the eyes of the interpreter who has deciphered the structure of the *Confessions*, Books I to IX, as arranged in a v-shaped schema, with the first half (Books I–IV) representing a descent away from God (*katastrophē*, in Plotinian terms) and the second half (Books VI–IX) narrating the return to God (the *epistrophē* of the Neo-Platonists), the two halves being thus split with precision in Book V (and, indeed, in the middle of this middle book, when Augustine leaves Carthage for Rome)—the interpreter who has seen all this has surely learned to recognize in this scheme all key events of sacred history: from the Garden of Eden in the guise of the garden of the stolen pears of Book II to Augustine's own exodus by means of crossing over the waters of the Mediterranean to the salvific reconciliation of the two wills, human and divine, in the agony at the garden of Milan which is also his Gethsemane. Yet, this interpretation would only take the reader as far as Book VIII, where one has reached the end of history as it was pronounced in the words "it is finished" (Jn. 19:30). Then, there is what can be aptly called the *eschatological* book (Book IX), as it is populated not only by a number of deaths (the biological deaths of Vervecundus, Nebridius, Adeodatus, and Monica, as well as Augustine's own symbolic death enacted by his immersion into the waters of the baptismal font), but also by the resurrection of the dead, as promised by the discovery of the incorruptible bodies of the martyrs Gervasius and Protasius (IX.7.16). The *Confessions* is a work that is neither contained within history nor content with reaching the end of history—it stretches even beyond that end, to the age to come. All the more, then, this reading leaves us even more perplexed with Book X (not to mention Books XI to XIII). Why does St. Augustine continue to write after the end of history? And what could possibly be the meaning of Book X, that is, in what kind of place are we to locate Book X, after the crescendo reached by the preceding books?

A Christian finds oneself living in two cities and between two ends: after Christ's coming in the flesh and before Christ's coming in glory. The enigmatic character of Book X of the *Confessions* (and hereby hangs, as we have seen, the problem of another classification) consists in finding Augustine precisely in this time: the time that is still of time and not yet of eternity. Book X constitutes a refusal, unique in its kind within Late Antiquity's proliferation of conversion narratives, to succumb to the temptation of seeing one's moment of conversion as the end of the road beyond which lies the unproblematic perpetuation of that moment's resoluteness. Augustine is not afraid to scandalize his reader when he writes *after* his conversion that he is still "a burden to [himself]" (X.28.39) or "an enigma to [himself]" (X.33.50) as he was before, or when he still writes, *after* his baptism, about his "sickness" (X.33.50) or his "wounded condition" (X.37.61). Does this language annul the difference that either baptism or conversion were supposed to make? If he is at the end of the *Confessions* the same man as he was at their beginning, what is his confession all about? What does he confess? Augustine's painfully sincere look allows us to glimpse at a theological difficulty that seems to have been overlooked: why history does not end with Christ's resurrection or ascension but it continues to drag until his second coming? What difference does it make living in this time-difference?

Such a question is symptomatic of the enthusiastic tendencies which Augustine prudently avoided. How much is at stake in the zeal that yearns to escape history can be thrown into a starker relief if we transpose the question about the ends of history to the beginning. To ask why history does not end when it seems to have reached its culmination is a question similar to that which asks after the reasons of history's beginnings. Why was there even history to begin with? Why does the Garden of Eden not coincide—not only symbolically but also temporally—with the Garden of Gethsemane? That is, why did God not do in the former what he *later* did in the latter? How could we explain or, worse, justify this "later"? But to ask this is ultimately a nonsensical question for, in a sense, it asks why creation had to be created—that is, it is a question that finds fault with the creaturely character of creation.

4

In the Between

"Evil becomes scandalous at the same time as it becomes historical."[1]

I. Between the beginning and the end

Let us begin again with the question about the beginning. At the end of the last chapter we posed the following question: why the Garden of Eden does not coincide—not only symbolically but also temporally—with the Garden of Gethsemane? That is why God did not do in the former what he *later* did in the latter? How could we account for this "later"? That is, how could we explain the time difference between these two moments, this delay that is history itself?

These questions open for us once again the meaning of the nature of time and of history as such. In order to formulate an answer we are forced to take into consideration the Scriptural account of beginnings and, most importantly, the subsequent readings, receptions, and interpretations of the seminal text of Genesis by those Christian authors who established a typology between these two gardens. Among these interpretations, Augustine's recasting of the salvific history in terms of his personal story as recorded in the *Confessions*, namely his journey from the garden of Thagaste to the garden of Milan is only an example, albeit one of the most brilliant ones. The scope of our discussion does not allow us to enter into an exhaustive survey of all the relative literature, nor does it permit us to engage the scholarly debates of Biblical criticism, except to the extent that it is relevant to our analysis. Besides, we are interested in the questions themselves and not on the historical criticism of their texts.

Scholars tell us that the reception of the first chapters of Genesis by the tradition of Christian exegetes takes considerable creative liberties with the Scriptural text and that they read too much back into the account of Genesis by discovering there a paradisiacal perfection that the text itself does not support.[2] Yet, the influence of the idea of a Paradise lost can hardly be exaggerated. The question, therefore, remains all the same: for the account of Genesis means to communicate an event, the story of something that happened, of a transgression that resulted in fundamental change in the state of the first humans—and through them of all creation (Gen. 3:14; Rom. 8:22)—for they are now cursed to live a life of sorrow and pain (Gen. 3:16), and of labor (Gen. 3:17–18), until they "return to earth from which [they] were taken" (Gen. 3:19). The subsequent expulsion from Paradise (Gen. 3:23–4) solidifies that what has been known ever since as the "fall" was indeed the intention of the text.

Augustine's narrative of the stolen pears in the garden of his childhood (Book II) is undoubtedly modeled after the Scriptural story of Paradise.[3] One line suffices to establish the parallel between Adam and Augustine's younger self: "How like that servant of yours who fled from his Lord and hid in the shadows!" (II.6.14, referring to Gen. 3:8–10). Yet, Augustine's recasting of the Garden of Eden dispenses with the illusion of a primordial perfection, for there is no room here for a state of innocence or even of ignorance. Contrary to Greek ethics, evil for Augustine is not a mistaken choice, vice is not ignorance, and sin is not a category of epistemology that could be regulated and rectified by degrees of knowledge. It is not enough anymore to *know* the good in order to desire it and to follow it. Contrary to Gnostic cosmology, evil is not external.[4] There is no external suggestion or seduction: the voice that prompts him to action does not come from the outside (cf., the serpent in the garden), but from his own self. Augustine is fully aware, as he insists, that his action was evil. In fact he goes a step further—and this adds a whole new dimension on the problem of evil—for his theft lacked any reasonable motive; his transgression was "for no reason ... there was no motive for my malice except malice" (II.4.9). What is quite unprecedented here is the deliberate reaction against ancient metaphysics for which "every art and every inquiry, and similarly every action and pursuit, is thought to aim at some good" as the opening line of the *Nicomachean Ethics* testifies.[5] Not only there was no good that motivated Augustine's action in the garden of Thagaste, but not even what Aristotle would call the *apparent* good: "No, I mean more: my theft lacked even the sham, shadowy beauty with which even vice allures us" (II.6.12). To do the evil for evil's sake is to leave evil unaccountable, without so much of a cause or a because—to leave it groundless. And this is perhaps the best one could do. For any attempt to explain evil is not only doomed, but also dangerous insofar as it renders evil intelligible. Explaining evil amounts, in some way, to defending it—and this Augustine was not willing to do.

As grim as his view of humankind's state may seem (especially if one were to take into consideration his view of children, even infants, as evil[6]), it nevertheless *liberates* the sphere of human action from the illusory longing for a return to an original perfection, now lost as a result of the fall from Paradise, where will is unaffected by being and vice versa. Such nostalgia not only captivates human imagination and compels it to compare the allegedly perfect beginning with the present, which cannot but be found lacking, but it also renders us captives to the desire of regression to some mythical unity. Thus, our understanding of ourselves and of our world becomes entirely orientated toward the past, toward that beginning, the *archē*: it becomes *archeological.*

Through such archeological understanding, time and history cannot be evaluated as movements with a direction towards an end, but only as moving away from a beginning—a falling away and a falling apart (it is, therefore, of little importance whether the story of Paradise is not the only one or only the first of the decline narratives, taken together or apart from similar Scriptural stories, such as that of Cain's crime and Noah's deluge, or even similar non-Biblical narratives).[7]

Whether we take as our point of departure a primordial perfection or not is decisive for the ways we understand history—whether history as a whole is a movement of perfection or rather the ongoing process of degeneration. In the end, what is at stake is the very question of time. If humankind enjoyed at some distant past some sort of perfection, then, the best it could hope for is to return to it. However, such a return could only mean a negation of history and the undoing of time. Furthermore, taking the Genesis account of Paradise literally envisages a time that humanity existed without time and, therefore, sustains the false impression of the possibility of an atemporal human existence. On the other hand, the denial of an archeological Paradise runs the risk of annihilating any difference, however formal, between the creation of the world and the fall, allowing the two to coincide, as if the world, and along with it the humankind, were always fallen. In this latter case, it is the creator God who takes the blame, as in all variations of Gnostic cosmologies. To navigate between these two extreme positions, some conceptual subtlety is needed.

The idea of a primordial *catastrophe*—to employ Plotinus's term that means quite literally a downward turn, a fall—can be found in most cosmological accounts: the fight between Tiamat and Marduk in *Enuma Elish*; the fight between the titans and the Olympians as well as the creation of Prometheus and Pandora in Hesiod's epics; the dismembered body of a god that becomes the birthplace of humankind (Dionysus in Orphism) or of the world (Ymir in the *Poetic Edda*), and so on. Apart from some typological differences,[8] what they all have in common is their effort to construct an etiology for the presence of evil in the current state of the world by alluding to a singular event of cosmic proportions that takes place either before or after the world's creation. They all succumb to the temptation of explaining evil.

One way to explain the current state of affairs of a world afflicted by both the *malum metaphysicum* as well as the *malum poenae* is to suppose a primeval crime as a result of which the world comes about. The advantage of this view is its simplicity: there is evil in the world because the world is evil, because the world originated in evil.

> [E]vil for Gnosticism is an almost physical reality that infects man from outside. Evil is external. It is body, thing, world. And the soul has fallen into it. This exteriority of evil immediately furnishes the schema of some thing, of a substance that infects by contagion. The soul comes from "elsewhere," falls "here," and must return "there." The existential anguish which is at the root of Gnosticism is immediately situated in oriented space and time. The cosmos is a machine for damnation and salvation. Soteriology is cosmology.[9]

However, this explanation operates on the exclusion of a good and omnipotent Creator. Thus, the alternative calls for a simple transposition of the fall from a time "before" the world's generation to a time "after" its creation. Now the world can be taken to be the good creation of a good Creator, which, however, for one reason or another, went awry. The problem of this position (and the story of Genesis is read along similar lines) is that it raises more questions than it solves:

a) Evil still remains fundamentally unexplained. In a world created "very good" (Gen. 1:31) by a good and omnipotent God, why is there even the possibility of evil? (The stock answer of freewill needs to take into consideration the fifth point below inasmuch as will is itself a movement).

b) Secondly, how could a good God be rejected by man, if we assume (together with Plato and Aristotle) that once one knows the good could not reject it, otherwise it is not truly good?
c) Thirdly, history, as we now know it, becomes an unintentional eventuality imposed on God by humankind's actions. Is it possible that God could not have known that man would fall? And if he knew, but did nothing to prevent it, is he still a good God?[10]
d) Fourthly, historical existence is now not only the aftermath of a mistake, but also the means of a punishment (pedagogical or otherwise makes no difference). Can God collaborate with evil without undermining his goodness?
e) Finally, and as long as we speak of a fall, we understand this fall as falling away from a state that, by implication, must have been un-fallen, that is, a state of perfection. In such an un-fallen state, were there time, movement, and change? And if not (since in a perfect state no change could take place for what could be the purpose of motion in a state that lacks nothing?) then, how could have a paradisiacal stasis been moved at all? That is, from a logical point of view (akin to Zeno's paradoxes, I admit), if there was a time (of perfection) before the fall, then the fall could not have happened, because something perfect cannot become imperfect, neither by itself nor by another, for if imperfection were a possibility of perfection, then perfection is not perfect, which is absurd.

Between the fall that creates the world and the fall that disintegrates the world after it has been created, Origen holds a unique position by accepting both. He could do this because he subscribed to a singular theory that acknowledges two creations. "The whole argument, then, comes to this, that God has created two universal natures, a visible, that is, a bodily one, and an invisible one, which is incorporeal."[11] We found this doctrine again in Gregory who must have borrowed it from Origen.[12] It is Origen, however, who introduces the subtle distinction in the language of generation, applying the term γένεσις to the first creation and γέννησις to the second (the difference between the two terms is inaudible in Greek).[13] Similarly, there are two falls: the fall of incorporeality into bodily existence (for example, the pre-existing souls into their bodies), and the fall of the material world into the travails of history as recorded in Genesis. It is of importance that for that first fall Origen does not use the term *ktisis* or *demiourgia* (the standard terminology for creation), but rather the term *katabolē* (καταβολή) which, although well attested in Scriptures,[14] carries the same connotations as the Plotinian *katastrophē*, if not worse, as it literally translates as a "throwing down."

This "throwing down" that was the cause of the material world's becoming was itself caused by boredom (*koros*). We have already referred to this idea in our comments on Origen's protology in the prolegomena. It is time to see it here in further detail. To the second question that we raised above, namely, "how could a good God be rejected by man, if we assume (together with Plato and Aristotle) that once one knows the good could not reject it, otherwise it is not truly good?" Origen gave the surprising answer "because of boredom."

> The creation of all rational creatures consisted of mind bodiless and immaterial without any number or name so that they all formed a unity by reason of the identity of their essence and power and energy and by their union with and knowledge of God the Word; but that they were seized with weariness [*koros*] of the divine love and contemplation, and changed for the worse, each in proportion to his inclination in this direction; and that they took bodies . . .[15]

It is as if the imperfection of God was his very perfection that failed to entertain the rational natures which contemplated him (as per the fifth problem above). I leave aside for the moment the objection that boredom is a category which, as we have seen, presupposes time, and indeed the *passing* of time, which Origen would not admit in the first creation, stipulating that time is the result of the second fall. There is a lot in the background of this idea as well as in the language employed by Origen that borrows from the Platonic myth of the descent of the souls and their embodiment in *Phaedrus*. Origen even attempts an etymological explanation for the derivation of the world *psyche* that seems to illustrate his theory of a primeval boredom:

> [W]e must ask whether perhaps even the word soul, which in Greek is *psyche*, was not formed from *psychestai*, with the idea of growing cold after having been in a diviner and better state, and whether it was not derived from thence because the soul seems to have grown cold by the loss of its first natural and divine warmth and on the account have been placed in its present state with its present designation.[16]

I suspect that Gregory of Nyssa's concept of *epektasis*, that is, of an eschatological desire that is satisfied but never satiated, is offered as a corrective to Origen's theory. However, it is Maximus the Confessor who should be credited with the success of rectifying Origen's protology. Without mentioning him by name, Maximus launches a critique of the pre-existence of the soul in one of his explanations on difficult passages from the work of Gregory Nazianzen (*Ambiguum* 42). A number of points makes clear that his target is Origen himself.[17] For Maximus the pre-existence of the souls and their subsequent embodiment as a result, if not a punishment, διὰ τὴν προγεγενημένην τῶν ἀσωμάτων κακίαν[18] ("on account of an evil previously committed by the bodiless") is unacceptable on two counts: first, it renders the grandeur of the visible creation, through which, as he says, "God is silently proclaimed,"[19] the effect of sin; and secondly, it forces upon God, and against his will, an outcome that he had not foreseen.[20] Maximus understands that a God who is constrained against his will ("μὴ θέλων ἐτυραννήθη"[21]) by evil is very close to a Manichean dualism. So he writes, "Clearly, the doctrine of preexistence shares the same characteristic as those who posit two cosmic principles and then pit them against each other in battle."[22]

The most systematic articulation of Maximus's critique of Origenism is to be found in *Ambiguum* 7. There the primeval fall of the first creation is rejected for the sake of eschatology—that is, if creation rejected God, while still united with him, who is to say that this will not happen again at the end of times? In Maximus's words:

> But if, as they maintain, what happened was the inevitable result of certain conditions, so that rational beings were moved from their abode and remaining in what alone is ultimately desirable, and consequently were broken up and scattered into multiplicity, we must ask in no uncertain terms: what proof do they have? For if what they say is true, it necessarily follows that rational beings, when found in the same circumstances, will undergo the same changes *ad infinitum*. For anyone who through experience is able even once to spurn something will find no reason to cease from doing so for all eternity. And if rational beings are to be swept about in this way, and are to be without any hope for an immovable foundation of stability in the Beautiful, what more pitiful condition of existence could there possibly be?[23]

In this passage, Maximus raises precisely the question that has guided us from the beginning of this chapter, namely, the question about the difference between the beginning and the end. In other words, Maximus's criticism is specifically directed in what he rightly suspects as the background of Origen's error, which is nothing else than the *eternal return of the same*. An absolute symmetry between beginning and ending captives the human mind, especially one that has been formed by Greek philosophy a little too much. "For the end is always like the beginning"[24] writes Origen.

> Now if we interpret correctly the passage which Moses writes in the forefront of his book namely, "In the beginning God made the heaven and the earth," as referring to the beginning of the entire creation, it is appropriate that the end and consummation of all things should consist of a return to this beginning . . .[25]

Yet, if the end is nothing more than a full circle return to the beginning (was it not the circle a perfect shape for antiquity?), then the difference between *archē* and *telos* makes no difference. This difference, which for Origenism as well as for Platonism (pagan and Christian alike), is indifferent is nothing less than the deferral, the time-difference between the beginning and the end, that is, *history*. If the end were selfsame with the beginning, then what is the *reason* of the beginning to begin with? What is the reason of creation? Why does God create?

What about, then, the story of Paradise in Genesis? It should be clear that it is not possible to maintain a paradisiac perfection as an actual, historical state at the beginning of creation. Maximus, therefore, posits the fall "*as soon as* [the world] is created" (ἅμα τῷ γενέσθαι, *Ambiguum* 42, 1321B). This allows him to avoid the problems of a (historical) time of perfection, while distinguishing between creation as created and creation as fallen, or put differently, between creation as being and creation as the theater of human action. The story of the Garden of Eden is not descriptive but rather proleptic: it lets an echo of the end be heard in the beginning, as often in music one hears at the beginning of a composition a theme that will be developed only at the finale. It is, after all, for the sake of that end that the beginning begins. "For he who is initiated in the ineffable power of the resurrection has come to know the purpose for which God first established everything."[26]

II. Between God and not God

> What is god, or what is not god, or what is in between—
> What mortal says he has found it by searching the farthest limit?[27]
>
> Euripides, *Helen*, 1137–9

In Book III of the *Confessions*, Augustine reads Cicero and thus philosophy makes its first explicit appearance in the dramatic unfolding of history that the structure and the narrative of the *Confessions* represent. Yet, ironically, philosophy was found lacking in one particular respect: Christ's name.

> In the customary course of study I had discovered a book by an author called Cicero, whose language is almost universally admired, though not its inner spring. This book of his is called the *Hortensius* and contains an exhortation to philosophy. The book changed my way of feeling and the character of my prayers to you, O Lord, for under its influence my petitions and desires altered. (...) Only one consideration checked me in my ardent enthusiasm: *that the name of Christ did not occur there.*[28]

This is ironic because philosophy since its beginning is concerned with beginnings—it is the beginning as such that interest philosophy, the beginning is interesting (and hereby a paradox lies concealed, for the interesting is already and always in-between thus in a difference from the beginning that is not indifferent to us)—nevertheless, philosophy ever since its beginning is interested with beginnings and thus one would have expected, Augustine suggests, that the name of him "who is the beginning"[29] might have been spoken.

Perhaps philosophy, like the young Augustine of Book III, is blind to *that* beginning—the beginning which is "in the beginning"—for it thinks of it (and of him) as the in-between, and so exceptionally, unsurpassably, the *metaxu* par excellence, that is, as the mediator. For Christ is also interesting and more so as Christ—that is, as the one who came to be between beings (*inter-esse*) and in our midst (*entos humon*, cf., Luke 17:21).

For philosophy, however, things are otherwise: while recognizing and acknowledging the fundamental difference and thus the distance between "what is god" and "what is not god"—to return to the Euripidean epigram—nevertheless a different answer or name is given to "what is in between" these two poles: for the Greeks, as it is perhaps expected, was a Christ-less philosophy (thus, Augustine's disappointment).

This is to think, with Plato, of philosophy as the *between* the world of God (or gods) and the world of "what is not god" that transfers (hence the original meaning of translation) something from the one world to the other, thereby establishing a communication, if not a communion, between the two orders. This hermeneutical function of philosophy was thought as particularly suitable for a daemon—that is, for a being who, on account of being a demigod, partakes of both realms: the human and the divine, being strictly speaking neither. Diotima gives us a memorable account of philosophy's demonic features in Plato's *Symposium*:

> They are messengers who shuttle back and forth between the two, conveying prayer and sacrifice from men to gods, while to men they bring commands from the gods and gifts in return for sacrifices. Being in the middle of the two, they round out the whole and bind fast the all to all. Through them all divination passes, through them the art of priests in sacrifice and ritual, in enchantment, prophecy, and sorcery. Gods do not mix with men; they mingle and converse with us through spirits instead, whether we are awake or asleep.
>
> *Symposium*, 203a[30]

Interestingly, for Plato the daemonic force best suited for this mission was Eros—an Eros who, as Plato through Diotima describes him in the lines that follow the passage cited above, found its closest resemblance in the person of a philosopher, and indeed the philosopher per excellence, Socrates. We shall leave aside for now this moment where philosophy, hermeneutics, and erotics converge together into a single point, as it deserves its own proper treatment. Instead, we shall follow Diotima's axiom as cited above: "Gods do not mix with men." For the excursion to the idyllic landscapes of Greek philosophy has confirmed in the most explicit way the problematic of the absolute disparity between God and humans, and thus, by implication, it has opened for us the question of theological impossibility. For if theology is merely *our* discourse about God and nothing more, then it cannot make any pretenses to know its object—such "theology" has rightly be called an anthropomorphism[31] and a conceptual (self-)idolatry.[32] If, on the other hand, theology is our discourse about God on the basis of and in response to God's always prior address to humanity, an address that was initiated by God's self-revelation and that it unfolds as an invitation to a dialogical conversation, then the human logos about the divine Logos (theo-logy) cannot be anything else than *the logos of the Logos*: in other words, it is of paramount importance that the Fourth Gospel calls God's self-revelation "the Logos" (the Word) who was "in the beginning" and who was eternally "with God" and who "was God" (John 1:1); for only as logos, as word, as discourse—a discourse that proceeds from God and is God—can God's self-revelation ground theology, now properly so understood. Theology as *the logos of the Logos* points to a hermeneutical uniqueness of a truth—indeed, *the* Truth, "I am the truth" (John 14:6)—that offers itself as interpretation and, moreover, as a self-interpretation.[33] A truth that is hermeneutical through and through.

"What is god, or what is not god, or what is in between them"—that was the tragic poet's question that guides our reflections here. It is important to notice that the poet poses the question in terms of "what is god"—*die Gottefragen*—and what is not god. The question of humanism is raised only negatively—that is, incidentally, as if "what is god" were a more readily available question; as if one knew the answer to that question better and therefore "what is not god" needed only to be asked in light of that first question and its corresponding answer. Yet, does the question "what is not god" point immediately at man? What is that that is not god? It is not the god-less for sure, nor is it the ungodly. It is not even the "mortal" which the poet mentions in the next line. For "what is not god" is the mortal only in one aspect—that is, only in respect to god's immortality. What, then, "is not god"—especially if we assume, as the chorus in *Helen* does, that one should begin with the question of god (of what god is)—that is, we should begin with the beginning and from the beginning? What is *not* god is what god is *not*: namely, this *not*. *What is not god is not.*

And yet, this *not* can be seen, and even see itself, and it can be spoken of, and even speak of and for itself, in light of what-is-god. But precisely because it can do so, it can also easily forget that seeing itself and speaking of itself is made possible only in a light of a *logos* that does not belong to it. Therefore, this *not* can mistake itself as that light and that logos, that is, it can assume that is self-illuminated (*autophoton*) instead of the recipient of a light external and quite foreign to itself (*heterophoton*). It is in this mistaken assumption that the problematic of sin opens up.

III. Original sin as the refusal of the origin

The metaphysical beginning—that is, the preoccupation with beginnings seems to function as a way in which one could speak of sin without lifting the veil over sin's proper face (in fact, this manner of speaking is that veil itself). Here sin becomes philosophical—but is this not philosophy's sin? Is it not this *our* sin qua philosophers, namely that when it comes to sin all we could do is run back to the beginning and speak the language of being?

Language suggests a closer association of sin with ontology than what one might assumed. Perhaps, in sin the connection between the ontological and the ethical remains intact. Thus, we trace the origin of the word "guilty" back to the Latin adjective *sons, sontis*—with probable derivation from the verb *esse*, "to be." If, on the other hand, think for a moment along the path of the Greek language, then, we are presented with a vocabulary familiar from Greek tragedy that includes such terms as *hubris* and *hamartia*; even though, there is a term closer at our philosophical home, namely that of *aitia*—a term whose translation into the Latin as *causa*, and thereby into cause, Heidegger has rightly protested as misleading.[34] It would be closer to its Greek meaning if we were to think of *aitia* as, for example, in *aitiatikē*, that is, the *accusative* case ("das, was ein anderes *verschuldet*"). When the philosopher speaks of causes and causality, he stands at the tribunal of reason as the accuser. To call the cause of a becoming—since only that which be-comes is in need of a "be-cause"—is to accuse, to proclaim that which comes-to-be as guilty.

Nothing is more familiar to philosophy and to philosophers than accusations. The first, and therefore the oldest of them, is the accusation of being—in the double sense: of having been, of having come-to-be. Recall the wish "not to have been" (μὴ φῦναι) of the tragic hero.[35] Perhaps it was Anaximander who first adopted the language of crime and punishment in philosophy, as a surviving fragment of his thought (DK 1) seems to suggest. According to this fragment, Anaximander must have seen the world as an immense crime scene where the crime of being is continuously committed. "Things" are coming to being, and that means that a moment ago they were not. Being is not theirs, for it can be taken away from them, as indeed it will, "paying penance and being judged for their injustices, in accordance with the ordinance of time." In a work unpublished during his lifetime, a youthful Nietzsche imagines Anaximander addressing beings thus: "... why are you here? Your guilt, I see, causes you to tarry in your existence. With your death, you have to expiate it."[36] The first accusation, then, that philosophy investigates, is that of being itself, of a certain unlawfulness in our origin, and in front of such accusation, we stand all equally guilty—an original guilt.

So much with *aitia* then. However, one cannot think of *aitia* without conjuring up its twin concept, namely, *archē*. In Aristotle's *Metaphysics* the two terms occur almost always together; wisdom, after all, is the knowledge "of the first principles and causes" (τῶν πρώτων ἀρχῶν καὶ αἰτιῶν, A 928b9). So, once more the question of beginnings presents itself—*protai archai*—a word that thanks to its Latin translation as *principium* has come to mean not only the beginning, but also the principle. What does it mean, then, to have a beginning—an *archē*—to find oneself already under the principle of the *principium*—of the beginning?

The answer now offers itself: having a beginning means above all that one is not *anarchic*, that one recognizes oneself at the very least as a derivative, secondary being—a being whose very being, precisely because it had a beginning, cannot mistake itself as its own beginning, that is, as *autarchic*.

The refusal to acknowledge that everything one has and is, except sin, is given from God, that is, the refusal of one's origin becomes the origin of sin, namely pride. The condition "except sin" points to a poignant conclusion: if the only thing of which I can claim ownership is sin, then, the only thing of which I can be proud of is nothing else but my sin, and being proud of one's sin amounts to being proud of my pride. Here already one detects that sin is devoid of content, it is an empty structure that seeks self-affirmation by running amok, that is, by pointing back at its emptiness.[37] It is in this sense that one can say that the origin of sin is sin, and that sin begets sin, which is to say, in agreement with the scholastic tradition, that sin does not have a proper cause, but it is only a presuppositionless negation.[38]

This denial of the origin, the forgetfulness of the beginning, can be also expressed differently as the denial of mediation and intermediation. Under this concept, sin is the desire for immediacy in two senses: a) with regards to time—as that which, unable to suffer the passing of time, becomes averse to time itself and to history; and b) as that which, unable to wait for the Other, seeks to bring about what is to take place on its own and by itself, becoming averse to any form of dependency on the Other. Kafka must have had this in mind when he wrote:

> There are two main human sins [*Hauptsünden*] from which all the others derive: impatience and indolence. It was because of impatience that they were expelled from Paradise; it is because of indolence that they do not return. Yet perhaps there is only one major sin: impatience. Because of impatience they were expelled, because of impatience they do not return.[39]

The first man sinned insofar as he wanted to achieve *without waiting* what was promised to him anyway but in due time. When the serpent promises our ancestors that eating of the forbidden fruit will make them "like God" (Genesis, 3:5), he promises them in fact what has already been planned by God as humankind's destiny: its *theosis*. The object of sin here is neither different nor contrary to God's will; rather sin consists entirely not on what one desires but *how* does one desire it, that is, on this double immediacy. What our ancestors heard in the voice of the serpent was the seduction of the here and now.

It is in this sense that sin can have its beginning only at the very beginning, at the very inception of time, when one cannot measure time, or by faking such timelessness. Thus, it is necessary *not* to avoid the controversial position that affirms the existence of sin already *at* the beginning of history, dispelling any notions of primordial perfection. Even though one can argue that there was no moment in history that humankind enjoyed a sinless state of perfection, history does not begin with sin or even less because of it, as Kierkegaard, for example, seems to think. For him, "without sin there is no sexuality, and without sexuality, no history."[40] I believe that Kierkegaard thinks of sexuality, in particular of sexual difference, as the result of sin only with respect to man for whom, on account of his corporeal nature, sin could be thought of as having such an effect—otherwise, he would have to explain why sin did not result in sexual differentiation when angels sinned. Yet, he is right in asserting that an angel has no history. "Even if Michael had made a record of all the errands he had been sent on and performed," he writes, "this is nevertheless not his history."[41] Still, to make history the aftermath of sin, or worse, the means of its punishment and correction, creates those inescapable perplexities that we have already discussed at the beginning of this chapter. Instead, history, which was never for the Creator an afterthought, *even if man had not sinned*, is God's means of a perfection yet to be achieved, the means by which such a perfection can be reached, as we expect eschatologically.

In fact, I would like to suggest, taking my prompt from the *Confessions*, that time as a movement toward perfection is at the same time the means of sin's undoing. For sooner or later (that is, in time), the *posse peccare* has to crumble under the confession of the *non posse non peccare*—admit, that is, that its capability is nothing more than its incapacity to transcend itself; its freedom is only a condemnation to repeating itself. In time, that is, in that very repetition, the sinner recognizes himself as such and despairs over his sin.[42] But his desperation at this point is already the birth pangs of his salvation—for a sin recognized as sin constitutes already a moment *after* sin, that is, a sin already delivered at the hands of time, and the hands of time, following St. Irenaeus of Lyon, are nothing else than the hands of God.[43]

It is, therefore, a characteristic of sin as such to be devoid of continuity. By this I do not mean to deny the habitual nature of sin, but only to call attention to its mode of existence which is no other than that spasmodic repetition which Kierkegaard has called "the sudden."[44] Like a tic, sin repeats itself almost compulsorily—although its compulsion is but an illusion facilitated precisely by its lack of continuity and can be exposed as such through the lens of mediation. As a series of spasms do not make up a gesture, let alone a movement, so the repetition of sin over a given amount of time fails to make up the content it lacks. It fails to arrive somewhere for in a strange way sin does not move, it only jumps.[45] Thus, sooner or later (that is, in time and over time), sin, no matter how enticing its promises and phantasmagoric its fantasies, must yield to *boredom*. Origen had missed the point: boredom is not the cause of sin, as he speculated. Rather, boredom is sin's inescapable effect. In boredom, however, sin has left the timelessness of the sudden and has already become subjected to the mediation of the temporal, in other words, it can now look at itself and become aware of its emptiness—that is, sin becomes perceptible as sinful.

There is, therefore, no need to emphasize at this point that boredom (as well as anxiety) are categories available to us only *after* sin and, therefore, Origen's explanation of the creation of the world—a creation which for Origen is still understood in terms of a catastrophe—by appealing to the boredom [*koros*] that the souls had experienced in their eternal preexistence suffers from the grave mistake of conceiving boredom *before* the beginning, indeed as the reason for which creation began. The mistake is quite obvious: in a timeless state of existence boredom makes little or no sense.

It follows that that which makes sin perceptible as sinful, to the sinner himself first of all, and thus opens the way for repentance is mediation as, above all, the reflection that can be afforded only by distance in time (diachronicity), and, secondly, as intermediation through the relation with the Other (intersubjectivity) against whom I have sinned.[46] In fact, both of these parameters are articulated in the famous parable of the prodigal son.

> "There was a man who had two sons. The younger one said to his father, 'Father, give me my share of the estate' [δός μοι τὸ ἐπιβάλλον μέρος τῆς οὐσίας/da mihi portionem substantiae]. So he divided his property between them [διεῖλεν αὐτοῖς τὸν βίον/divisit illis substantiam]. Not long after that, the younger son got together all he had, set off for a distant country [εἰς χῶραν μακράν/in regionem longinquam] and there squandered his wealth [διεσκόρπισε τὴν οὐσίαν αὐτοῦ/dissipavit substantiam suam] in wild living."
>
> Lk. 15:11–14[47]

"Father, give me my share of the estate." What did the prodigal son ask for? His share of the estate, says the English text, translating a Greek word with many and important meanings: give me, says the Greek text, "μέρος τῆς οὐσίας"—give me, that is, a part of *being*, give me part of your paternal *essence*.[48] What the prodigal son asks for is, in a deeper sense, impossible. Either he asks for a share in the father's fatherhood, he asks to become himself his own father, to give birth to himself, or, he asks that *being itself* be given to him, that his being might be given. Yet, what sense does it make for one to ask to be given that which one already is? "Give me my being, father" says that being who is a being given. That life is given it can mean different things at different circumstances: it can either mean that life as given is a gift over which one is full with gratitude or, it can mean that life is *a* given, something over which I have no choice, and therefore no freedom. Life itself, insofar as it is given, it constricts and constrains my freedom. Thus the given-ness of life can give rise to two opposing experiences: on the one hand, that of gratitude and thanksgiving, and, on the other, to an experience of imprisonment from which one desperately seeks to free oneself, seeking an alleged independence.

What the prodigal son really asks for in asking for his share of "the estate"—even if we were to understand "the estate" according to the language of the parable in purely pragmatic and practical terms, as referring to his property—is nothing more than his freedom. The very figure of his father is a reminder that everything he has, his life included, is not his, but something for which he ought to feel indebted to someone else.

And a man who feels indebted, that is, under debt, cannot presumably feel free. It is freedom that the prodigal son longs for, the kind of freedom one understands under the prized word of "independence." He simply wants to be independent. Even such independence is bought on the price of a being that is divided, of a division engrafted in the heart of being. "Διεῖλεν αὐτοῖς τὸν βίον"—the verb διαλαμβάνω means to divide, to separate, to cut off, but also to grasp separately.[49] The passage, then, speaks of a life seized, claimed and, as such, separated and torn into pieces. The prodigal son's life is not anymore the life of his father but his own. Like Oedipus in Thebes—to anticipate the discussion of our next chapter—he replaces his father by becoming the father of himself and thus fatherless.

As soon as he has left the paternal house, he finds himself in a "distant land." But what land can be distant from the father who holds all in his hand? Is there a place devoid of God? To look back at Augustine's series of perplexities with which he chooses to open his *Confessions*: "No, my God, I would not exist, I would not be at all, were you not in me. Or should I say, rather, that I should not exist if I were not in you, from whom are all things, through whom are all things, in whom are all things?" (I.2.2, p. 40). "The distant land" is nothing more than *the land of distance*, a land in which everything is presented in some distance, no matter how close we come to people and to things. It is a world that is fragmented—into so many fragments as many people and places as are there; fragmented into gender and race, into different languages and ethnicities, into different social classes, separated from each other and ultimately from ourselves.

In such a land, one cannot but "squander his wealth"—again, the Greek text is helpful: "καὶ ἐκεῖ διεσκόρπισεν τὴν οὐσίαν αὐτοῦ": there one *scatters* one's life and oneself; in this land, which we call our world, one cannot but be a scattered being. As scattered, we desire our unity, we desire to be one, to be *whole*. First, to be one with ourselves, for not unlike that demon who, when questioned about his name, answered "legion" (Mk. 5:9), *each of us is many*. That's why the return to the Father begins with a return to oneself: "εἰς ἑαυτὸν δὲ ἐλθὼν/ in se autem reversus" (Lk. 15:17) which we translate "when he came to his senses" even though a more faithful translation would have rendered it "having returned to himself." Indeed, one can be distant not only from others, but from oneself; and the return to the Other takes place by a turning toward oneself. For no matter how much we try, no matter how much we love each other, our efforts remain frustrated, our intimacy with one another never quite succeeds in making us one. Hence that sense of failure, of a desire unsatisfied and unsatisfiable that awaits us at the end of our efforts. The land of distance is not so much a land defined by geography, but rather by a way of being, our way of life, which is to say that we do not reside in the land of distance, but rather we have become *it*: each in his and her way making out of ourselves, as Augustine says, a wasteland ("et factus sum mihi region egestatis," II.10.18).

Sin cannot be conceived *in abstracto*, that is, outside and without the relation between the sinner and the Other to whom the sinner is accountable. That Other, on the other hand, cannot be the sinner himself, as if that to which the sinner is accountable could be his own sense of duty, "the master [who] lies in us,"[50] or his conscience. Such a misinterpretation would have returned us to such notions as that of autarchy and autonomy. Furthermore, there is not such thing as a consciousness, in the strictest phenomenological sense, which could stand alone without being in reference to and in relation with an Other. "[T]he *inescapability of the other* is undeniable."[51] No, first and foremost sin presupposes the very relation it is understood to have damaged; it presupposes the dialogue that sin has interrupted. The essential background of sin, therefore, without which the concept of sin would have been cancelled out, is what Paul Ricoeur has called "the fundamental situation of a man who finds himself implicated in the initiative taken by someone who, on his side, is essentially turned toward man; . . . [namely] a god who is anthropotropic—before being anthropomorphic."[52] That is why, for Ricoeur, "sin is a religious dimension before being ethical; it is not the transgression of an abstract rule—of a value—but the violation of a personal bond."[53] A phenomenological description of sin, therefore, cannot but begin by focusing on the structure of being-before-God, of being-in-relation, or to use just one word, of being a *prosopon*. I have analyzed elsewhere the phenomenology of *prosopon*,[54] and so there is no need to return to that analysis here. It would be sufficient to ask: "if sin is primarily the rupture of a relation,"[55] then how can we best envision the character of that relation?

Nietzsche, in his *Genealogy of Morals*, puts forward a creative interpretation of guilt (*Schuld*) that traces its origins back to debt (*Schuld*).[56] The most fundamental of debts is, according to him, that which is owed to one's ancestors—it is, in other words, an ancestral debt, being indebted for one's origin, an indebtedness that requires from the members of the tribe a repayment in the form of "sacrifices . . . feasts, music, honors; above all, obedience . . ." Nietzsche's analysis concludes with the hypothesis that the first gods, the very concept of divinity, emerged out of such contractual relationship with one's ancestors—invested now for the first time with divine attributes on account of one's fear for them.[57] History keeps adding to the original, ancestral debt interest to such an extent that it reaches a symbolic proportion that is beyond repayment. Eventually, only a god would be able to pay back what is owed to god, a god that would become man in order to sacrifice himself in satisfaction of humanity's debt. That is, for Nietzsche, the "stroke of genius on the part of Christianity"—namely that

> God himself sacrifices himself for the guilt of mankind, God himself makes payment to himself, God as the only being who can redeem man from what become unredeemable for man himself—the creditor sacrifices himself for his debtor, out of *love* (can one credit that?), out of love for his debtor![58]

Taking a step that anticipates Freud, Nietzsche argues that, at some time in history, the relationship between two individuals, namely creditor and debtor, was internalized within a single individual.[59] Of course, as long as that contractual relationship was external and intersubjective did not constitute guilt *as such*. It became that, when indebtedness and correspondingly the feeling of "being-indebted-to," that is, the feeling of an unsettled account, of unfinished business, is transferred within oneself, becoming thus intrasubjective.

It is precisely this line of thought that Heidegger picks up in the famous analysis of sections 57 and 58 from *Being and Time*, where the silent call that conscience itself is pronounces only one verdict and always the same: "guilty."[60]

Heidegger's idea of life as "a series of repayments on a loan that you did not agree to, with ever-increasing interest that will cost you your life;" that is, life as "a mort-gage"[61] takes us back to Anaximander's fragment. Yet, in the time between Anaximander and Heidegger a decisive transformation took place to which philosophy ought to pay attention: I speak of the almost imperceptible change from the neutral "the one" (τὸ ἕν) to the masculine "the One" (ὁ ἕνας), that is, of the movement from impersonal being to a *personal* God. In Greek the difference is mostly in the definitive article and in English it is inaudible, but it is a difference big enough to make all the difference in our discussion for, by being in relation with a personal God, being must also be understood as personal, and between persons the primary relation goes under the name of either *eros* or *agape*. It is telling that in the prophetic utterance sin takes the form of an accusation for *adultery*.[62] In turning away from God to other self-made gods, and offering to them what one ought to offer only to God, not only we are proven unfaithful, but also we have shown ourselves guilty of a certain promiscuity. Thus the other side of adultery is, sensibly enough, *idolatry*.[63] In the prophetic language adultery becomes the most eloquent metaphor for idolatry, yet idolatry itself can, in turn, become a metaphor for sin in general. After all, as we have seen, sin is nothing else than the positing of the self as its own self-made idol. Thus, the Great Canon, which is sung in the Orthodox Church during the last week of Lent, puts such an anguished confession in the mouth of the faithful with great clarity: αὐτείδωλον ἐγενόμην, "I have become an idol to myself" (*Great Canon*, Ode 4th). Self-idolization becomes, therefore, the final accomplishment on a course that began with the denial of beginning only in order to end in the self-denial of one's end.

Let us bring these reflections on sin to a conclusion by returning to the beginning—even though we have advised against the allure of the origin (and of Origen). What Paul Ricoeur called "the scandal of evil" is precisely that sin which has been known since Augustine as "original." What is scandalous is not sin as such, but rather its qualification as *original*. However, to understand the concept of original sin—especially in its late, doctrinal formulation—as some biological category is a misunderstanding. To say, as Paul does in Romans 5:12, that "all men have sinned" in Adam, does not establish a chronology or a chronicle of sin, whereof Adam is, quite unaccountably, the first sinner. Rather, it suggests the "transbiological and transhistorical solidarity of sin"[64] and "the communal dimension of sin."[65] If it is true to say that I sin "in Adam," then *it is equally true that Adam sins in me and because of me*. Thus, Kierkegaard's enigmatic remark that Adam and every individual is "at once himself and the race"[66] becomes intelligible.

> The Adamic myth reveals at the same time this mysterious aspect of evil, namely, that if any one of us initiates evil, inaugurates it—something Pelagius saw very well—each of us also *discovers evil*, finds it already there, in himself, outside himself, and before himself. For every consciousness which awakens when responsibility is taken, evil is *already there*.[67]

The notion of original sin, therefore, underscores the universality of human responsibility. Insofar as I do not constitute an irruption in time, a moment of a new beginning without history, then the evil committed by my ancestors is also my responsibility—a theme around which Greek tragedy is structured. Yet, this is not a responsibility mediated by the notion of inheritance—as we have seen, sin is without continuity, it cannot be located in time, even though time "dislocates" it. Thus, ancestral sin is not before or behind me but contemporaneous with my sin. It is this synchronicity of evil, or rather the timelessness of each act of evil, that calls me to assume responsibility for all evil—for outside of time evil knows of no individuation, yet it is what individuates me (for, as we have seen earlier in this chapter, the only thing I was not given, the only thing of which I can take pride is my sin). To repeat Kierkegaard, I am at once myself and everyone (the human nature, the race, etc.). "There is only one salvation for you," Father Zosima is reported to have said:

> Take yourself up, and make yourself responsible for all the sins of men. For indeed it is so, my friend, and the moment you make yourself sincerely responsible for everything and everyone, you will see at once that it is really so, that it is you who are guilty on behalf of all and for all.[68]

5

After Evil

"After all this the question always remains: Will the evil end and how? Does creation have a final purpose at all, and, if this is so, why is it not reached immediately, why does what is perfect not exist right from the beginning?"[1]

These questions conclude Schelling's seminal essay on the *Essence of Human Freedom*, a philosophical investigation on the age-old problem of theodicy, namely, of the justification of an omnipotent and benevolent God in face of the reality of evil. Confronted with such a dilemma one might attempt either to deny evil, and thus allow evil to triumph, for there is hardly anything more evil than willfully turning one's eyes away from it, or to concede that God himself is evil, as in the dualism of Manichaeism. Neither option is, of course, philosophically or theologically satisfactory. To the question *unde sit malum*, Christian metaphysics, since Augustine, has sought an answer by both adopting the Neoplatonic notion of evil as lack and deficiency (*privatio boni*) and by affirming the freedom of human will. The problem, however, is thus masked rather than resolved: as Augustine himself came to realize during the Pelagian controversy, human will is not that free after all for, although it can choose what is evil (*posse peccare*), it cannot choose not to choose it (*non posse non peccare*). Nor can it be assumed that human will was free originally (that is, in its prelapsarian state), for choosing what is evil, before or after the fall makes no difference, is already evil and thus moral evil remains, as to its origin, unexplained. From the viewpoint of culpability restricted to personal agency alone, neither original sin, as it was discussed in the previous chapter, nor the tragedian's notions of a "guiltless guilt" and hereditary curse make sense. Thus, the root of evil and, by extension, evil itself, remains shrouded in mystery.

Schelling was well aware of that difficulty.[2] He understood—and here lies the ingenuity of his insights in the *Freiheitschrift*—that theodicy cannot be understood apart from *teleology*. It is far from accidental that in the series of questions with which we opened this chapter the problem of evil is raised simultaneously with that of history, its direction and its duration ("does creation have a purpose and, if so, why is it not reached *immediately*? Why does not the perfect exist *already* from the beginning?). These questions ask after nothing less than the existence of time itself. And rightly so, for, as I would like to argue in the following pages, the problem of evil and therefore of theodicy is resolved only once evil is seen as *temporal*—perhaps in both senses of the word, as something that exists as a moment in time and as something that exists only for the time being.

Schelling is not alone in raising the question of evil's end. We get a glimpse of such an eschatological vision in Plato's *Lysis* when Socrates rather surprisingly, asks the question "whether evil [τὸ κακόν] is eliminated . . .?" What follows is a description of a paradisiacal perfection:

> I wonder, if evil is eliminated, whether it will be possible to be hungry or thirsty or anything like that. Or if there will be hunger as long as human beings and other animals exist, but it won't do harm. Thirst, too, and all the other desires, but they won't be bad because evil will have been abolished.[3]

The hypothesis is almost immediately dismissed as ridiculous [γελοῖον τὸ ἐρώτημα]. Yet (self-) ridicule is not the last word. For, with a daring second step, Socrates leaves such a possibility open: "Who knows?"[4]

I. From Eden to Thebes: philosophy in the age of Greek tragedy

In order to illustrate the effects of time on evil, I have chosen to engage in a sustained reading of Sophocles' drama *Oedipus Rex*, with a particular question in mind: what is the connection of evil and time? My contention is that ethics all too often has avoided the temporal intricacies that lie on its foundations. The contemporaneity of two given choices with which one is supposedly presented in taking a decision, or the—unthought, I would claim—distinction between a before and an after of the ethical act, testify to the inadequacy with which ethics, old and new, have approached the problem of time. Sophocles provides us with a text that most appropriately, as I will attempt to show, raises this very question insofar as Oedipus's story presupposes a horizon of time without which the drama remains distorted and unresolved. Under the unifying gaze of a-temporality (the *nunc stans* facilitated and even provoked by the unity of time—exemplary in Aristotle's eyes[5]—of the play's dramatic unfolding on stage) Oedipus's responsibility remains incommensurable to his fault that is ultimately not his. Thus one is left only with a lesson in fatalism, which of course cannot explain the transformation that Oedipus undergoes in Sophocles' sequel, *Oedipus at Colonus*.[6] The resolution of the drama, its turning point, relies exclusively on this small detail: on the realization that this is not Oedipus's first time in Thebes, that he has been there *before*. The irony of the drama—its hidden core from which it continues to excite an irresistible allure—is the fact that this so crucial *before* could have never been brought to view the *first* time around, and thus, in some essential way, Thebes is a city that one can never visit for the first time, but it is rather a city to which one can only and always return.

Like Charles Ryder in the opening chapter of *Brideshead Revisited*, Oedipus knows himself only when he utters these words: "I have been here *before*."[7] The allusion to Waugh's novel is far from superficial insofar as *Brideshead Revisited*'s leitmotivs, sin and grace, make it a thinly disguised re-casting of St. Augustine's own narration of sin and redemption in his *Confessions*.[8] What these three literary examples share in common is a conception of ethics as temporal insofar as in all three it is time that brings about the character's self-realization (*anagnorēsis*).[9]

For the *Confessions* to be confessed, Augustine must place himself simultaneously in the position of the narrator as well as in the position of the subject narrated. By doing so, however, he is far from coinciding or overlapping with himself as it might have been expected, but rather, he maintains a continuous distance of himself from himself, for if the narrator were to overlap with his narration there would be no story to tell (except the very act of narrating itself). Therefore, confessing the *Confessions* depends on that distance of oneself from oneself, a distance occasioned by nothing else than time. Passing judgment on oneself and one's own actions is made possible only from the point of view of a *self revisited*—through and thanks to time.[10]

For Oedipus to know who he truly is he must first know what he has done. But neither patricide is patricide nor incest is incest as long as Oedipus is not Oedipus. Seen from the perspective of the present, that is, from a timeless perspective, the murder of Laios is little more than an accident, as his relationship to Jocasta is as normal as any relationship between a man and a woman. It is only when one looks at these relationships *in time*, from the perspective of past and future, that one understands them fully.

One could draw two guiding conclusions from these initial observations: a) that synchronicity knows of no ethics and b) that at the moment of the act's present one cannot ascribe an ethical value to that act (as if there were intrinsically good or bad acts[11]). The question is, then, *when* can one pass judgment? The answer can only be one: at the end. It is this that the audience are told at the final scene of *Oedipus Rex* which, as many other tragedies, ends with the proverbial admonition not to praise one's happiness before the end,[12] and indeed the absolute end: be it that of one's life or that of history. With these thoughts we turn now to the Sophoclean drama, with the intention to read the story of the son who kills his father as philosophically parallel to the story of the father who kills his son as told in Kierkegaard's *Fear and Trembling*.

II. Evil's refusal of the temporal

Sophocles makes the plague the incentive moment of his play. A plague has befallen the city and the elders have gathered outside the royal palace to ask for the intersession of their king. Of course, a plague is not *any* disease: it is an infectious disease—that is a disease that manifests itself not through the single individual (for it does not affect so much the individual *qua* individual) but individuals—in the plural—insofar as one belongs to a community and precisely *on account of* that community. The plague is a sickness of the community, of the city (that is, the polis). In fact, one could say that the plague is a *political* disease. (In that sense, I believe, it was also used in Camus's novel by the same title.)

Further, the dialogue about a certain sickness reflects a sickness in the dialogue. Reading carefully the exchange that takes place between the chorus and Oedipus one comes to realize that it has only the semblance of dialogue, for it proves to be no more than an echoing monologue. Oedipus speaks but what he says returns to him, as his own reflection. The chorus says nothing that Oedipus does not already know (as he admits, 58) and in fact he has already done what the chorus came to suggest (41–3).

Moreover, there is that reflexivity of sickness between city and king on which Sophocles insists (by repeating the same word three times within the space of two verses). Thebes is sick but "no one is more sick," as he declares (59–61), than he himself—this reciprocity stops short only before revealing the real nature of that relationship, that is, if Thebes is sick that is *because* Oedipus is sick. Finally, there is the "*homou*" of verse 64 that brings city and king together into a solid amalgamation, the result of which is a Theban Oedipus and an Oedipal Thebes. Oedipus's sickness is, of course, not that of the plague—for *that* Oedipus knows of no community, he has no other and therefore he is the individual *singulare tantum*—his is the sickness of (his own) identity and I mean the sickness inscribed in his very name—for Oedipus's name is, quite literally, the name of disease ("swollen feet"). Sophocles connects these two diseases—the disease of the individual (i.e., the self-enclosedness in sameness) and the disease of the communal (that has projected evil outside itself in some mythical Other, in this case, the Sphinx) and makes the one the manifestation of the other. If, in other words, Oedipus infects Thebes with his disease this is because Thebes, this Thebes, is established on the Oedipal expulsion of evil, the exclusion effected by Oedipus solving the Sphinx's enigma but, conversely, if Oedipus is sick this is because he cannot trace his identity to a time beyond his triumph—a triumph of reason—over the enigma posed by the Sphinx.

Both King and polis are formed on the same twofold basis: the exclusion of otherness and, what comes as a result of this, the self-enclosure within sameness. Nowhere is this better illustrated than in Oedipus's encounter with the Sphinx. If the chorus has gathered to ask for Oedipus's help, and if Oedipus is so determined to provide it, this is because he rescued the city from the ancient monster. For both the city and its King the eradication of that monstrosity (temporal in any way, for the Sphinx returns as the plague) is a point of reference that determines who they are—this city is precisely a city liberated of the enigmatic presence of the Sphinx and Oedipus is he who solved her enigma. The Sphinx stood at the borders of Thebes and thus she had defined the very boundaries between the city and what was other than the city—the city's other. That topological function was also transposed to another level: the Sphinx functioned as the symbolic presence of evil, very much like the serpent in the Garden of Eden.[13] By threatening with death anyone who would prove unable to solve her enigma—the enigma that she herself was—she also stood as a reminder of that ultimate limit of death. By solving her puzzle, Oedipus had not only explained a riddle, but had also defended the city against the evil other. In doing so, he becomes a new founder of this Thebes where what is evil is identified as what is other (the foreigner, immigrant, outsider, etc.). The forgetfulness of evil lies in the constitution of the polis. It is precisely this new identity that Oedipus evokes in his opening address and it is of this identity that the chorus reminds him as they plead *that he remains the same*. This telling request comes from the priest's lips (53) and expresses a veiled anxiety for change—the same anxiety that Oedipus would try to ward off by declaring, when it would be too late, that "nothing can make me other than I am" (1084–5). The exclusion of otherness is completed with the imperative that one remains the same, in recognition, perhaps, that otherness and evil might lie ultimate within oneself. "Not everything in man is human."[14] In his insightful reading of *Oedipus Rex*, Goux draws a comparison between the tripartite monster that Oedipus confronts (the Sphinx) and the tripartite structure of the soul as presented by Plato's psychology.[15] This allows him to be critical of Hegel's reading of the confrontation between the Theban King and the Sphinx as the inaugural moment of rationality.[16] If this is so, Goux argues, it is on the expense of the irascible and concupiscent elements of the soul that the exclusive identification of man with reason does not defeat, but rather allows them to operate on their own, it leaves them uncontrolled, and ultimately enables them to return as the desire to kill the father (irascible) and sleep with the mother (concupiscent). The Freudian reading of the story of Oedipus is based precisely on such unconscious subversion. The Egyptian element, as represented by the Sphinx, "has become internal and individual."[17]

In that recurring sameness, bereft of the possibility of opening up to what is other, time runs as a continuous present. The enigma of the Sphinx, as we shall see later, had opened up for the city the horizon of time—in fact it was nothing less than an invitation to think man's existence as stretching from past to future, to think the present as a present offered by the past for the sake of a future and to think the one in terms of the other, to see, that is, the past under the light of the future. Having solved the enigma of time, Oedipus closed his being and that of the city in the present—and thus they live, as Creon aptly describes (130–1), trapped in the now. Asked as to why the city did not investigate the death of Laius, he says "we were compelled to attend to instant needs" that is to an instantaneous present that unfolds into a series of "nows" without ever forming a memory, that is, a past. It is precisely from the past that both Oedipus and Thebes are alienated ("we were compelled to let slide the dim past")—and it is this very past, in the form of a plague—that returns now in order to upset the present.

The "dim past" that the city was compelled to let slide is more specifically spoken of in Greek as τἀφανῆ, that is, the things invisible. Oedipus's answer in the next line is one of those celebrated moments of intentional ambiguity that the Sophoclean irony builds on: ἐγώ φανῶ; this can mean either "I will reveal [these invisible things]" that is, "I will make apparent what now remains unapparent," but it can also mean "I will be revealed." In fact, *Oedipus Rex* does both: what the future reveals is precisely the past, and Oedipus is shown as being equally implicated in both. One could, then, speak of two "Oedipuses": the Oedipus that lives in the present and for the present, oblivious of the past and without a future, and the Oedipus of the future when he will come to realize his past. It is such a double Oedipus that the great soliloquy of verses 216–75 presupposes, if it is to be effective. The Oedipus of the present curses the Oedipus-to-come, an Oedipus ignorant of himself turns against an Oedipus gifted with the knowledge of his ignorance. One needs to notice here that Sophocles makes the former—Oedipus of the present—nothing more than a figment of Oedipus's imagination, while the latter—the Oedipus to come—is bestowed with reality.[18]

There is a moment in the play where these two "Oedipuses" meet each other, and therefore a moment when, proleptically, the future meets the present. Teiresias enters the stage exactly as Oedipus will later exit it: blind, led by a child, knowing the truth. Already in the beginning, the audience is given a fast-forward glimpse of the end. Teiresias is the future of Oedipus. As such, as Oedipus's future, he has the power to effect the first break on the solid surface of Oedipus's imaginary, self-created identity.

That break occurs precisely at verse 437 when Oedipus asks: "who are my parents?" Oedipus's question is about his past, but this past cannot be recognized unless it returns, in the guise of a future, to his present. The question about his beginnings will signal his end and thus the promise of a new beginning.

Oedipus introduces himself as soon as verse 6 of our text with the phrase: "Oedipus, renowned to all." Apart from a certain sense of pride, this self-proclamation conceals the real problem, that is precisely that of his identity, insofar as Oedipus is "renowned to all" but unknown to himself. The reply of the priest (31–8) gives us a glimpse of Oedipus's self-established identity: Oedipus is the city's redeemer. The reference here is to the Sphinx. As we saw, the farther back into his past that Oedipus can go at this stage, the oldest memory that he can recall of himself is his triumph over the Sphinx's enigma. That episode becomes the birth-moment of his imaginary identity. This is the very moment when Oedipus gives birth to himself—it is no accident that immediately afterwards he would replace his father, quite literally, by taking his father's place on his mother's bed, becoming as it were, his own father.

Oedipus, at this point, is only he who was able to solve the Sphinx's enigma. What distinguishes him, in other words, from the rest of the people and makes him to be "the best among men" (46) is precisely his ability *to think*. Oedipus is not a Hercules. He does not labor with physical strength. Nor is he an Odysseus, winning the day by his personal skill and persuasion. Oedipus is solely a thinker. But what kind of a thinker? The answer lies in the enigma. The Sphinx asks:

> What is that animal that
> In the morning walks on the four,
> At noon on two feet
> And at the evening on three?

Oedipus's answer was only a word: *man*. A word that, strangely, referred back to him, insofar as he *was* his answer. In retorting against the monstrosity of the Sphinx with the humanity of man, one could say that Oedipus becomes the first humanist, and in doing so solely with confidence in his reason, he becomes the prototype of a rationalist too.[19] If, however, Oedipus succeed in solving the Sphinx's puzzle this is because he thought of a particular characteristic of man and his existence: he conceived of man in his *timely* manifestation—of the human being as projected in the horizon of time, for it is precisely time that the three periods of the day—morning, noon and evening—indicate. Oedipus thinks of man as that being that persists in temporality—he thinks, in other words, of *being and time*. Oedipus gives the ancient monster (the monster of antiquity?) a modern answer and thus he fittingly becomes not only father of himself but also father of that fatherless epoch that is called modernity. Oedipus has given birth not only to himself but to the mind—to himself through the mind. Oedipus's answer means this: *I think* (for I am able to think the enigma that I am) *therefore I am*.

> *Cogito ergo sum* has to be understood as a victory cry; the shout of the successful "patricide," the exultation of the son who now knows that he no longer depends

> upon ancestors or anyone else in order to stand on his own two feet. (. . .) Descartes's patricidal gesture is incommensurable, in its import and its radicality, with any individual killing of an individual father. Descartes is a principial and abstract Oedipus. He denies the ontological dimension of paternity.[20]

Oedipus, the philosopher. The Sphinx was nothing else by Oedipus's own, at once, irrational and evil element, brought outside of him and made other than him in order precisely to avoid recognizing it as his own.[21] Such unrecognized evil, however, remains inescapable (or, if one prefers, unforgiven). Indeed, not only it returns in the form of the plague, but also it returns with the urgency of a responsibility that demands to be assumed insofar as it has been made now public.

An indication of that self-made identity was given at the beginning of the drama when Oedipus names himself. If he can give himself his name that is because he has first given himself to himself. As with our names, our identities are given by others. To receive a name is to recognize that you have a beginning (the two events almost coincide, the giving of a baby's name follows shortly after the baby's birth). By refusing to receive a name, Oedipus indicates his refusal to accept his beginnings—for that we would have to wait until the very end of the drama, for at the end his beginning will be revealed and accepted. It is only at the end that the beginning can be revealed or recognized. For now, however, Oedipus remains "anarchic" i.e., without *archē*, in a semblance of eternity, for without beginning he is also without end, that is, without boundaries. This unlimited self, a self that knows no limits or limitations, cannot but be an imaginary self.

Oedipus, as an infant, is exposed to exteriority, an exteriority symbolized first by his expulsion from the paternal house and, furthermore, by being left on Mt. Cithaeron to die (1170–80). This exposure constitutes undoubtedly a traumatic event of which he has no memory and yet, in a way, he is constantly reminded by his very name. His name, after all, is the name of that trauma in the literal sense ("Oedipus" means "the one with swollen feet"). His self-made identity can be explained as an imaginary construction in which he finds rescue from that painful exposure. His identity is imaginary insofar as it denies the trauma of his infancy. It is an imaginary identity because it does not take account of the beginning (*archē*) and thus it fakes anarchy. Therefore, Oedipus's identity is determined by two parameters: expulsion of otherness and self-insulation in sameness. It is this and nothing more that the Delphic oracle says, translating it as it were in the language of a double crime, when it prophesized that Oedipus will kill his father and sleep with his mother. Patricide, that is, violence directed toward the Other, as well as incest, that is, the pathological attachment to the Same, is what Oedipus has always committed by being who he is. Oedipus's fatal encounter with his father at the crossroads outside Thebes meant to signify that every murder is potentially a patricide for every Other is always the parental Other—the Other who gives birth to me since I cannot be my own origin. Similarly, every returning upon oneself (*cor curvum in se*) constitutes a form of incest. Jocasta is for Oedipus an Other bereft of otherness; an Other upon whom I have projected my own familiar image, i.e., my expectations, my feelings, myself. Such "Other" is nothing but a mirror that reflects back my own reflection and therefore my relationship with such an "Other" is deeply incestuous. The recognition of this double crime will break down the imaginary self and will "give birth" to a new identity: "this day will give you birth and kill you" (438).

From this perspective, the actual "carrying out" of his double crime is irrelevant, for, in a sense, patricide and incest have been committed long before they take place and when they occur, they are only symptomatic—a delayed manifestation, so to speak—of Oedipus's patricidal and incestuous identity. It is crucial to recall that Oedipus executes both acts *precisely* as a result of his desperate, and, by all accounts, sincere effort to avoid them. It would be mistaken to read the inescapable crime in terms of some fatalism. Oedipus cannot avoid the double crime as much as he cannot avoid who he is—and if he *is* the crime that he commits, even prior to committing it, this is precisely because he wants to avoid being who he is.

In fact, it is only after he has fulfilled the Delphic prophesy that Oedipus can escape his criminality. Thanks to the double crime, Oedipus's identity is now both externalized and realized and thus the possibility has opened up for him to recognize himself. It is curious but telling that Sophocles is not interested in the crime itself but only in its recognition; that is why he sets both plays, *Oedipus Rex* and *Oedipus at Colonus*, after these events. No wonder, then, that the only person who can see Oedipus's relationships properly is the seer Teiresias who, precisely by being blind to the present, can see what remains for most invisible (τἀφανῆ), that is, the future and the past.

Oedipus's knowledge of himself depends on his knowledge of others, for example, he will fully know who he is only once he comes to know that his wife is also his mother. However, we have seen how Oedipus invented himself on the basis of the exclusion of otherness. The first step toward recognition of otherness takes place through a double confrontation, first with Teiresias and then with Creon. The irruption of hostility at this stage is to be counted as a positive development for it signals, for the first time in the drama, Oedipus's awareness of the otherness of the Other. So far Oedipus had refused to enter into any relationship with another—even that of hostility. In doing so, he had refused himself the very possibility of relationship insofar as the first step of a genuine relation with the Other cannot but be hostile. If the Other is truly Other and not another me, he or she appears as *other* than me and therefore my first recognition of the Other cannot be anything else but oppositional. At this initial stage, I still have the option to express my hostility towards the other (i.e., fear, scorn, hatred, disgust, indifference, etc.) and thus acknowledge his otherness, even if it is in these negative terms, or avoid hostility for the sake of badly-understood "respect" that, in the name of tolerance and acceptance, creates a homogeneous community without otherness and thus without the possibility of relation (i.e., the image of the plague that befalls the city in the opening of the tragedy). Between these two options, the refusal of hostility is worse for, by assuming for itself the good prematurely, it can lead nowhere but must remain stagnant in inauthenticity. In the words of G.K. Chesterton: "The next best thing to really loving a fellow creature is really hating him . . . The desire to murder him is at least an acknowledgment that he is alive."[22] Under this light, "love thy neighbor" (Matt. 12:13) is the same commandment as "love thy enemy" (Matt. 5:44) *for the first enemy is the neighbor*, until love transforms that very enemy into a neighbor, truly so understood.

III. After evil: the good

There is an old story, preserved for us by Xenophon in his *Memorabilia*, that relates how an adolescent Hercules, having reached a fork in his path, is asked to make a choice between two different ways: the way of virtue and the way of vice. The story underlines the idea of an ethical choice. There are moments, the story seems to say, when we have to choose, and our choice is ultimately between vice and virtue, between what is good and what is evil. This idea is, of course, all-pervasive.[23] It is a notion that has stayed with us from early Christian literature (*Didache*) to contemporary cinematic epics. Philosophically, it posits the possibility of good and evil as two autonomous, independent from each other, and contemporaneous choices.

In Xenophon's story, Hercules (the ethical agent) is imagined as somehow standing *before* these two ways and therefore *outside* them. The question, then, is: what is the ethical status of Hercules before he makes his choice? The story would seem to suggest that he is neither on the path to goodness nor on his way to wickedness—he is ethically nowhere. Our difficulty is now becoming more explicit, for a subject who is no-where is also (ethically at least) no-one. There is no neutral category (neither good nor evil) that one could ascribe to Hercules before his character-determining choice. One, therefore, has to conclude that he is *not*—that is, it is the choice between the two paths itself that brings him to life. But how can no-one choose? Would it not make more sense to say that Hercules does not choose the path that he will take but rather (and much like Oedipus) he is being chosen?

Alternatively, one could read the path on which Hercules finds himself before his choice as being a mixture of both good and evil—after all it is that path that leads to a bifurcation between these two. In this case, though, the radicality of the difference between good and evil vanishes. If he has trod (read, tried) both of them, then his choice on the fork of his way is superfluous. If the course of Hercules is nothing but a symbol of the course of one's life, it is difficult to see what could justify such an ethical bifurcation at a later and seemingly arbitrary stage. Why then? Why not earlier or later? Perhaps this fork on the road is symbolic of an important decision—one that would have some bearing on Hercules' life. Even so, one feels the aporia of how one is to choose—that is, how one ever knows the way of the good from the way of the evil? How one knows without presupposing a subject who has an absolute knowledge, a subject for whom the outcome of each way is presented in a totalizing panorama?

Perhaps, then, one never knows, because one cannot know. That is after all what makes an ethical choice ethical—that one has to choose without such foreknowledge. But then what is one to make of the bifurcation between good and evil, the splitting of the road into two distinctive and different paths?

An ethical choice is never a bifurcation of exteriority—the story makes little sense if taken literally, for the course of one's life cannot "split" into two lives;[24] instead the moral dualism is always experienced as a split internalized in the subject itself: schizophrenia within me. What accounts for this pathology is the false positing of the good alongside with the evil, the fact that one takes the good as simply being the opposite of what is evil and furthermore the fact that one places these two opposites side by side—presenting thus oneself with a dilemma.

But no such dilemma between good and evil can be maintained. To posit the good alongside the evil equates the one with the other, as if the good were no different than the evil except insofar as it is good while the evil is evil. It is precisely this mistake that allows for the danger of ethical relativism, that is, of the undoing of ethics, for it does not take long before one starts asking what makes the good good or by whose standards is it good. The good is not of the same nature as the evil and therefore it has nothing to do with it. If the evil is simply the privation of the good (an Augustinian theme), how can one place it next to the good without imagining that one could also place a night next to a day and so on? It is never, then, the case that the good can be presented as an option or as a possibility alongside the evil, for the two can never exist *contemporaneously*. Only a Manichean system could allot evil equal status with the good. Furthermore, an evil that co-exists with the good is an evil that limits the good. The question then is: *what good is that good for*? A good that is contemporaneous with evil is no good at all.

Augustine, writing precisely against such a Manichean conception of ethics, comes to the bold conclusion that good and evil are not two independent phenomena but rather the good is evil's self-realization as such in time. Evil is a moment in the temporal unfolding of the good. What allows Augustine to arrive at this groundbreaking conclusion, demonstrated by the very narrative of his *Confessions*, is his assumption that sin precedes freedom. For Augustine one is free because one sins and not the other way around, where freedom might have been affirmed as freedom towards or from sin. *Original* sin means simply that sin is just that: original, prior and antecedent. That is, inescapable. The famous story of the stolen pears tells indeed that "I sin because I can." The reader is, however, soon to realize that the freedom indicated by this *posse peccare* is itself un-free, for were its affirmation to have any validity, were its "yes" to have any power, it should have also the ability to say "no"—*posse non peccare*. Such is, though, a possibility that remains humanly impossible and therefore the "freedom" of sin cannot free itself from the very sin through which it was initially manifested. Thus we meet Augustine in the garden at Milan tortured by the realization that what had begun in the first garden as "I sin because I can" was proven to be rather an "I cannot not sin." This realization, however, cannot be given *unless* one goes through the suffocating depths of the first; in fact it is nothing more than the impasse of the first. It is its truth. In other words, the hidden truth of *posse peccare* (the illusion of freedom) is the *non posse non peccare* ("freedom" as illusion). But this truth is revealed only in and through time; for example, through the repetition of sin. What does this truth constitute? A sin that is aware of itself as sin is already beyond sinfulness, an evil that judges itself as evil cannot do this but in the name of the good, by becoming good. Therefore, left to itself—and to the unfolding of time—evil undoes itself by turning into its very opposite, by becoming recognized as good.[25] Between, then, evil, as represented by the first garden of the *Confessions* (an allusion to Eden), and goodness, as symbolized by the second garden (an allusion to Gethsemane), the difference is only time.[26] The distance between Thebes and Colonus is a distance measured in years, not miles. The difference between good and evil is before anything else a *time difference* and good and evil are primarily temporal categories. Or, to say the same thing differently, past and future carry an ethical weight.

Where classical (i.e., *static*) ethics fails is precisely in its failure to take *time* into account. Leaving time outside the ethical picture results in all sort of impossibilities, like the ones we have examined above, and makes ethics ultimately untenable. Recall that this was precisely the point where Oedipus had also failed and that the Sophoclean drama resolved itself by opening its protagonist to the horizon of time. It is only within this horizon that one can distinguish between a beginning and an end and therefore allow ethics to play out *as the difference between the two*. Oedipus at the beginning waits, like Adam, for "his eyes to open" (Gen. 3:7)—of course, Sophocles' irony lies in the fact that the opening of Oedipus's eyes would only come at the moment that he blinds himself—that is, at the end, as with Adam who begins to live at the very moment of becoming mortal. In *Oedipus Rex* death precedes evil: the death of Laios precedes the evil that befalls the city and which is nothing else but the denial of that death. Evil is finally overcome only once Oedipus recognizes death—not only the death of the father but his own death, for in giving birth to himself he has given birth to a stillborn; it is this recognition that opens up the possibility for life and for the forgiveness at Colonus.

In the end, it is life itself that at the beginning is revealed or rather perceived as death. It is the good itself that at the beginning is considered as evil. The imaginary self, trapped as it is in the illusion of self-existence and in the illusion of freedom which is nothing else than non-existence, cannot perceive the life-giving exteriority but as death—as a mortal trauma from which it must retreat, which it must avoid and repress. However, that which at the beginning appeared traumatic and fatal, at the end will bring the promise of a new life. Evil is, then, nothing more than the denial of life's trauma and as such, the nostalgia for non-existence, that is, the nostalgia for a timeless existence, which, since it was never given or experienced, means nothing else but the nostalgia for nothing itself.

The Second Garden: *Being Divided*

6

Will and Grace

I. Time and narrative in *Regione Dissimilitudinis*

The confessional aspect of Augustine's *Confessions*, when by confession one understands primarily praise and prayer, calls attention to what is almost destined to fail with regards to God, namely, language. The problematic of language is introduced by Augustine's emphasis on the literal meaning of infancy, as his "unspeaking stage" in life (I.6.10, 44). Language comes as a remedy to a fundamental experience of a primordial separation, a separation that manifests itself first in the distinction within/without, inside/outside: "I was frustrated in this, because my desires were inside me, while other people were outside and could by no effort of understanding enter my mind" (quia illae [voluntates] *intus* erat, *foris* autem illi, I.6.8, 44). The first experience of fallenness is the separation from others, in terms of a distance in space.[1] To this one, immediately, Augustine adds a second one: the separation from oneself, in terms of a distance in time: "My infancy has been so long dead now, whereas I am alive" (I.6.9, 44), as he states at the beginning of his attempt to remember the immemorial. "For what is it that I am trying to say, Lord, except that I do not know whence I came into this life that is but a dying . . . I do not know where I came from" (I.6.7, 43). The mystery of memory, as it will be fully developed in Book X, is first disclosed through its opposite: forgetfulness. "So I have been told, and I believe it on the strength of what we see other babies doing, for I do not remember doing it myself" (I.6.8, 43).

Both experiences, separation from others, namely space, and separation from oneself, namely time, are but two aspects of the same condition, of the *diastemic* nature of the fallen creation.[2] Language, then, is grounded in such a condition and it is itself regulated by it. In Book X language is a privileged example of time, for to speak takes time and one cannot speak but in time. But language also "takes" space: the meaning of a word is determined by its position in a sentence. And even though one can use language to communicate with another, not every other speaks my language, as young Augustine learns from his painful attempts to master Greek (I.13.20, 52). Even among speakers of the same language, language can be the source of misconceptions and misunderstandings, thereby my separation from the Other is emphasized all the more, even while attempting to reach out and communicate oneself to another.

If separation exists, however, it is because of the fragmentation that the self undergoes in falling into what Augustine will later call, borrowing an expression from Plato's *Statesman*, "the region of dissimilarity" (*regione dissimilitudinis*, VII.10.8) and, therefore, of multiplicity. Here things are always many, for we have turned away from the One and therefore lost our ability to be united. The temptation to read into this description Plotinus's *katastrophē* is understandable, but it seems that Augustine is thinking more in Scriptural terms. "And when you pour yourself out over us, you do not lie there spilt but raise us up; you are not scattered, but gather us together" (I.3.3, 40). The gathering-together, this *synaxis* points already to a theme which Augustine reserves for the end of his *Confessions* namely, the Church. God called the people and the nations that were scattered over the face of the earth into the unity of his Church when he "poured [himself] out over us" in the pouring-out of the Holy Spirit in Pentecost. And like the Holy Spirit hovers over the amorphous abyss during creation, giving unity to what we call for that very reason a *universe* (see XIII, 5.6, 345–6), so also at the historical re-creation of the creation, the Holy Spirit unites once more and brings together those who were dispersed because of their pride, their *superbia* as narrated in the story of the tower of Babel. I believe we can read echoes of that event back in the opening lines of Book I, in "the proof that you thwart the proud" (I.1.1, 39). For both, the dispersion into multiplicity as well as the Pentecostal gathering into unity, are underscored by *language*. In the former, God divides human language, in the latter that division is overcome so that when Peter preaches to the nations, each is able to understand him in their own language (Acts, 2). Language, then, has in the *Confessions* a double function: as a symptom of separation and as a factor of unity. The ambiguity of language will be analyzed further in the forthcoming discussion.

For Augustine, the overcoming of the self's fragmentation is achieved by his conversion which now could also be understood more accurately as a process of *universion*, that is, of turning toward the One and, as a result, becoming one with oneself and with others.

On the thither side of the *diastematic* wall,[3] in the absence of separation, nothing is everything. God is everywhere because he is nowhere in particular. Yet he is fully immanent because fully transcendent. The two distinctions that cause multiplicity in this "region of dissimilarity," that is, the distinction of one sensible thing from another in space and time, but also the distinction of one suprasensible thing from another by virtue of difference and otherness, do not apply to God, from whom, every thing, whether sensible or suprasensible, draws its unity insofar as everything is, in some respect, one. In the One beyond space and time and beyond difference and otherness no distinction can be maintained. As a result, two important characteristics of God emerge, his simplicity and, in consequence of that, his ability to permeate all things without being any of them (cf., God's omnipresence).

Nevertheless, language remains *diastematic*, which means that, when man attempts to apply it to God, to speak *of* God instead of speaking to him, language breaks down into a series of paradoxes like those Augustine records in I.4.4. At the end of these aporiae he asks: "After saying all that, what have we said, my God, my life, my holy sweetness? What does anyone who speaks of you really say?" In agreement with the apophatic tradition of theology, Augustine understands that God's incomprehensibility is the only concept by which we could approach God.[4] Thus, he exhorts his reader "to find by not finding rather than by finding fail to find [God]" (I.6.10, 45).

The polarity between distance and proximity as introduced in Book I is further elaborated in Book II. There Augustine portrays himself as, at once, Adam and prodigal son. The language that follows suit these two metaphors. The opening phrase *recordare volo* has as its object a series of anecdotal episodes in which the adolescent Augustine finds himself. Their common denominator is summed up in the phrase *dum ab uno te aversus in multa evanui* (II.1.1). The "dispersion" mentioned in the same sentence is made exact by the "turning" from the one to the many, a turning that has as a result the evacuation of one's very self, leaving oneself empty. This emptiness frames Book II from its beginning to its end, where Augustine, choosing the Biblical image of the Prodigal Son, describes himself as a wasteland: *et factus sum mihi region egestatis* (II.10.18). By doing so, Augustine draws the best out of the parable of the prodigal son for, unlike the Gospel story, he does not *come* to a distant land—that land is for him now, more interior than his inmost part, *interior intimo meo* (III.6.11), but rather, in a twist worth only of a genius, he *becomes* that land.

The emptiness of the self is the inescapable result of the self's attempt to ground itself or, better yet, to ground itself on itself, to become its own ground. Even if one would like to stand on his own feet, he would still have to stand on some ground (metaphysical or otherwise) other than oneself. As with the failed attempt to autobiography, which is thwarted by the paradox of a self-narrated self without beginning (cf., chapter 3), so too the self-grounded self must be at the end groundless. "But to whom am I telling this story . . . and why?" asks Augustine. His answer is telling: "So that whoever reads [these writings of mine] may reflect with me on the depths [*de quam profundo*] from which we must cry to you" (II.3.5, 64). A confession that is *de profundis*: not only earnest, but also emerging from the abyss one has fallen into or the abyss one has become (*in imo abyssi*, II.4.9). "I leaped down from your strong support into destruction" (II.4.9, 68).[5] Thus the human deep calls on to the divine deep, *abyssus abyssum vocat* (Ps. 41/42:8).

In the first paragraph of Book II, Augustine's gaze is turned equally toward himself, his life, but also toward the narrative of that life: the recollection of the *recordari* must first passed through the cohesiveness of the *colligens*; for a scattered self cannot be recollected in the unity that every narrative constitutes. Ultimately, there is not a story of a scattered and fragmented self as such because no such story could be told. If the episodic and fragmented story of Augustine's life is to be told, it must be rhapsodized into a unity—a unity, however, that can only come from a unified self. How is the self unified? As we have mentioned already, by a process of *universion*, by its return to the one God. What accomplishes the unification of a divided self? The entire narrative of the *Confessions* is the answer to this question. For the *Confessions* not only records the story of Augustine's conversion, but also illustrates the means by which this conversion became possible. The recollection of a self by means of narrative, a narrative of the ways in which the self endured over time, allows oneself to become one. To say it differently, the answer that the *Confessions* offers to the problem of a self-divided self is *time* and *narrative*.[6] So, paradoxically, that which *at the beginning* appears as the effects, if not the causes themselves, of division, *at the end*—but only at the end—they are proven to be the means of division's overcoming. The ambiguity of language is resolved by time, and the ambiguity of time is resolved by narrative. Ricoeur's epigrammatic statement that "we are following, therefore, the destiny of a prefigured time that becomes a refigured time through the meditation of a configured time"[7] finds an exemplary application in Augustine's *Confessions*.

Between the first and the second garden of the *Confessions*, that is, in the distance from the pear-tree of Thagaste to the fig-tree of Milan, a distance that, incidentally, covers the entirety of human history, from its beginnings in the Garden of Eden to its eschatological consummation at the Garden of Gethsemane, the will, the problem of the will, will as a problem has come to full relief.[8] Thus we read in Book VIII of a division more fundamental than that between the subject's interiority and the exteriority of the world:

> How did this [monstrous] situation arise, how develop? The mind commands the body and is instantly obeyed; the mind commands itself, and meets with resistance. When the mind orders the hand to move, so smooth is the compliance that command can scarcely be distinguished from execution; yet the mind is mind, while the hand is body. When the mind issues its command that the mind itself should will something (and the mind so commanded is no other than itself), it fails to do so. How did this [monstrous] situation arise, how develop? As I say, the mind commands itself to will something: it would not be giving the order if it did not want this thing, yet it does not do what it commands.
>
> VIII.9.21, 201[9]

The monstrosity of which St. Augustine speaks here has to do with the realization that more fundamental and anterior to any mind/body dualism is a dichotomy that takes place within the self. The mind's resistance to its own commands is indicative of a duality that, in the context of his text, becomes articulated in the language of *two wills*. "There are two wills, then, and neither is the whole: what one has the other lacks" (VIII.9.21, 201). However, how can we ascribe two wills to the same subject without rendering it asunder? And if the will is only a faculty of the mind, would it not the positing of two wills force us to confront the absurdity of two minds and thus two subjects?[10] Yet, St. Augustine's existential experience of the strife between two opposing wills hints at a more fundamental idea beyond all these apparent absurdities: it reveals a duplicity that is an essential structure of the human being, the division implicit in Kierkegaard's enigmatic remark that "[a]t every moment, the individual is both himself and the race."[11]

II. Freedom beyond freedom

We have seen how Kierkegaard's enigmatic assertion played a role in the explication of the conundrum posed by the notion of original sin (cf. chapter 4). Phenomenology can helps us understand better the experience of the self's self-division since it has called attention to the formal discrepancy between the I and consciousness itself, thus rendering the I both superficial in its function and artificial in its origin (Sartre[12]) or, to say the same thing in a different language, it reminded us of the substantial difference between one's nature or being and the particularized and individualized existence of that nature in the hypostatic existent (Levinas[13])—a non-coincidence into which, in the words of Paul Ricoeur, "a certain problematic of the unconscious can be introduced."[14] It is precisely this difference within oneself that makes a first experience of *freedom* possible, for the human person can never conform to nature by default—for even if he or she does so, it will have to be his or her decision. *Personhood cannot be reduced to being or existence.* That is, I am not a being or even my being. The self I am, this I that says "I" is in struggle over against being. Better yet: the I is (exists, but also is what it is) only insofar as it differentiates itself from (its own) being. To have said that "I am (my) being" could have only two meanings: either the impossible declaration that I am *the* Being, impossible not only in its megalomaniac arrogance but also because *the* Being cannot speak in the first person, cannot be *a* being, or, alternatively, the I that utters these words, by means of saying what it says, sabotages itself, commits a metaphysical suicide of sorts, and returns to the anonymity of being. To say "I am my being" is to say "I am no-one and no-thing;" it is to say "there is no I."

It was Jean-Paul Sartre who first declared the ego to be an essentially *artificial* and *superficial* invention of the consciousness itself:

> Perhaps, in reality, the essential function of the ego is not so much theoretical as practical. We have noticed, indeed, that it does not bind up the unity of phenomena; that it is limited to reflecting an *ideal* unity, whereas the real and concrete unity has long been effected. But perhaps the essential role of the ego is to mask from consciousness its very spontaneity.[15]

The question we need to ask is this: for what reason does the ego need to play this essential role of masking from consciousness its very spontaneity, that is, of hiding from consciousness what consciousness is, its very nature and essence? The answer is clear: consciousness cannot bear its own freedom, it cannot stand itself. It always seeks to escape itself, either by its engagement with the world (that is the function that intentionality had played in the unreflected level, as Sartre notices, "[b]y intentionality consciousness transcends itself. It unifies itself *by escaping from itself*"[16]) or by erecting the ego, in front of whose restrictive function consciousness can now cast itself, relieved from its own unlimitedness. Hence Sartre's enigmatic remark: "[c]onsciousness is frightened by its own spontaneity because it senses this spontaneity as *beyond* freedom"[17] Yet, "spontaneity" is here just another word for freedom. Thus we have arrived at this enigma: what does a freedom *beyond freedom* could possibly mean?

Beyond freedom indicates a freedom that cancels itself out by being precisely free—so free that is "restrained" by its very inability and impossibility of being un-free.[18] A freedom needs limits and restrictions in order to be perceived or perceive itself as free; that means that freedom is only a freedom towards this or that ("freedom to . . ." or "freedom from . . ."). Without limits, freedom falls upon itself, it becomes its own monstrous "limit"—that is, the very lack of any limit. Thus, the ego saves freedom (from itself) by bringing it back to a place where there are restrictions and limitations and therefore, freedom can continue to be free by being limited.

Consciousness's anxiety on the face of its own spontaneity—what is known elsewhere as an existential vertigo or nausea—is underscored *by the redoubling of the will.* Sartre defines the will as "an object which constitutes itself for and by this spontaneity,"[19] over which the will is powerless. "The will directs itself," Sartre continues, "upon states, upon emotions, or upon things, but it never turns back upon [the willing] consciousness."[20] *The consciousness that says "I will" is not the willing consciousness.* Thus, two wills emerge—the will of the consciousness itself, for to the extent that *willing* is an intention, it is of the intending consciousness, and the will of the I, erected by that same consciousness. We are here reminded of Augustine's anguish in saying: "I was the one who wanted to follow that course, and I was the one who wanted not to. (...) I was at odds with myself and fragmenting myself" (VIII.10.22, 202). It is now essential to see why these two wills necessarily oppose each other.

Consciousness as absolute[21] seems to allow no room for an Other, and less so for the wholly Other. In fact, it assumes for itself those attributes that theology traditionally assigns to God. Thus, consciousness is a cause of itself;[22] unlimited and thus infinite;[23] its representations are eternal,[24] and so on. There is, here, the repetition of the old philosophical blasphemy that identifies seeing and knowing, *theaomai* and *theoria*, and attributes both to God (*theos*), but also to the philosopher's ability to know (we shall return to this theme in our discussion of the narcissistic self in the next chapter). Consciousness, like a small god, can indeed regard everything from its point of view without been seen.[25] In spite of such god-like characteristics, we have witness—and this *is* the witness that consciousness brings against itself, the "testimony of our conscience" (2 Cor. 1:12)—a consciousness ready to cast itself in front of whomever will promise to free itself from itself, that is, from its freedom. Such is the promise fulfilled by the ego. Yet, as we have seen, the ego is only a superficial and artificial invention of the consciousness. What for a moment appeared as the promise of a saving god, now it becomes nothing more but a false idol. The ego is the golden calf that consciousness, in a tellingly idolatrous moment, erects for itself.[26]

In the invention of the *ego*, consciousness implicitly recognizes itself as un-free in its unlimited freedom or unable to endure such a freedom. Its struggle and misery, the inevitable state of being-at-odds-with-itself, comes precisely from knowing that it itself has no master and that it itself cannot completely fool itself in believing that it—namely, its ego, the I—is its own master.[27] Its will is not—and now it should become evident that it also *cannot* be—the same as the will of its ego. At the end of this analysis, the *ego* emerges as the process and, at the same time, as the end-result of an on-going self-deception, of fooling oneself into believing to one's self-mastery or autonomy.

Thus, the ego for phenomenology (or the super-ego for psychoanalysis) is not so much a false god, or rather it is a false god only insofar as it is a false, imaginary Other.[28] Here the theme of epistemological idolatry is linked to its ethical counterpart of self-idolatry or, as it is known in the Scriptures, of adultery (cf., chapter 4). For by the self-deception of its autonomy, consciousness hides its radically heteronomic structure, that is, it hides from itself or it forgets its own constitution. What gives itself to itself is precisely a givenness anterior than any self-constitution.

Of the fundamental bifurcation of the self we hear also in Bergson's *Time and Free Will.* In his critique of the spatialization of time, Bergson discovers a superficial self and a deep-seated self. He calls the former "the parasitic self" and the latter "the fundamental self."[29] What is of crucial importance, however, is that for Bergson the differentiation between these two aspects of the selfsame self occurs thanks to and by means of *language*.

> Consciousness, goaded by an insatiable desire to separate, substitutes the symbol for the reality, or perceives the reality only through the symbol. As the self thus refracted, and thereby broken to pieces, is much better adapted to the requirements of social life in general and *language in particular*, consciousness prefers it, and gradually loses sight of the fundamental self.[30]

The ego is then the ego of language and of the symbolic order. If it can be loved (the possibility and origin of the narcissistic self) this can be done only once the ego has been formed or informed by language. We shall see later that desire is never mute but always articulated as a demand, that is, meditated by language.[31] It is for this reason that the imaginary follows after the symbolic (which is also to say that in the order of the *Confessions*, Book III on the imaginary self follows after Books I and II on language).

We only need to add to the above what Sartre calls "the profound irrationality of the notion of an ego."[32] For in the ego itself "there is always a ground of unintelligibility."[33] Here, and contrary to the wishes of the phenomenologist, we meet Freud's idea of a ground of irrationality with which our conscious life is somehow complicit.[34]

III. A will wide asunder

It is generally and rightly credited to St. Augustine the insight to recognize in the faculty of the will a force that was, unlike the ethical theories of the Greeks, independent from the mind—a force with its own life, so to speak, a life whose first biographer St. Augustine becomes by the way of his *Confessions*. It is in fact in the *Confessions* that both the will's independence from knowledge and, therefore, its arbitrariness are for the first time in the history of ideas affirmed. For without the teleological nexus of knowledge the will could neither be explained nor justified. The problem of the will confronts us with a certain irrationality in the depths of our existence. The confrontation with this problem in Book VIII of the *Confessions* is modeled after a scriptural text taken from Romans 7:

> I do not know what I do. For I don't do that which I will, but I do that which I hate. And if I do what I don't want to do, I agree that the law is good. As it is, it is no longer myself, but the sin that dwells in me who does it. I know that nothing good lives in me, that is, in my flesh. For I find the will with me, but doing what is good I find not. For what I do is not the good I want to do; no, the evil I don't want to do—this I keep on doing. Now, if I do what I don't want to do, it is no longer I, but the sin that dwells in me who does it.
>
> Romans 7:16–20[35]

It was this very passage that set St. Augustine on his long and arduous way of re-examining his anti-Manichean affirmation of a free will.[36] The agony expressed in this passage, even to the point of paroxysm, reflected best in the very syntax of the text, is indicative of the experience of finding oneself at odds with oneself. The crux of the problem lies in the second half of verse 18. We do possess the ability to will (what, confusingly, is called "free will") but already this will is determined to do "that which I hate," i.e., sin (therefore, my will is not free after all). My inability not-to-sin (*non posse non peccare*) is not the result of ignorance, for I *know* the good that I want, I even *want* it, but of my will's inability to execute it ("but doing what is good I find not"). The name that St. Paul gives to this inability is "sin." Here "sin" refers primarily to the cause and only subsequently to the result. It is "sin" that causes me to sin. Sin begets sin. To put it differently, if I sin, if I am able to sin at present or with regards to this or that, that is because I have already sinned. "The sin that dwells in me" already prefigures the notion of *original sin*, that is, a sin ancestral of sin, the first sin by which all subsequent sins and sinning itself was made possible.[37]

The experience, however, described in the passage quoted from the *Epistle to the Romans* is the perplexity one finds oneself in by realizing that what one supposed to be one and whole is in fact perceived by the self itself as two, as the splitting into two, *a division that continuously takes place within oneself*. We need to be careful at this point: the passage speaks at the same time of both a unity and a division within that unity. If the division was not within the unity of oneself there would have been no perplexity, indeed, no pain. It is not, in other words, a battle between two individuals that concerns us here, nor is it a battle between two distinct, separate forces or realities. Rather what we have here is a self that becomes aware of itself as that division between two aspects that are equally itself. St. Paul speaks, for example, of two laws, one seen in "his members," the other in his "mind" or "after the inward man." Both, however, are his, they are him.

How could we otherwise explain man's differentiation from his will which this passage describes, if the will becomes one and undifferentiated with man's nature? And does it not this differentiation constitute my freedom, a freedom either *for* sin or *from* sin? Is the freedom of a human being to be understood with respect to his nature, to his will, or both? Whence is human freedom? Our experience of our fallen condition makes hardly possible to talk of a freedom according to nature: I cannot chose to eat, since I cannot chose not to eat without suffering the consequences of my nature—there is, indeed, a certain violence in nature.[38] I can choose what to eat (relative freedom of choice), but I can also choose to abstain from food (e.g., fasting), sleep (e.g., vigil), and every other biological compulsion that my nature imposes on me. If the will is to be understood only as a natural activity, how can I resist the compulsion of the will itself? How can I ascend to the other's will in saying "your will be done"? Could a nature oppose what is natural and could nature fight against nature? Whence is human freedom? On the other hand, if one is to be identified completely with one's nature, then why do you punish them or award them? To ascribe the will to nature, without reference to the human person, would threaten the entire spiritual ethos of the Church.

No other literary passage has illustrated the theory of two wills more memorably than the eighth book of the *Confessions* that narrates the famous moment of St. Augustine's conversion at the garden in Milan. In his own rendering of Romans 7:16–25, St. Augustine writes:

> To find my delight in your law as far as my inmost self was concerned was of no profit to me when a different law in my bodily members was warring against the law of my mind, imprisoning me under the law of sin which held sway in my lower self. For the law of sin is that brute force of habit whereby the mind is dragged along and held fast against its will, and deservedly so because it slipped into the habit willingly. In my wretched state, who was there to free me from this death-doomed body, save your grace through Jesus Christ our Lord?[39]

Corresponding to the two selves are an equal number of *voluntates*, two wills fighting each other: "the old and the new, the one carnal the other spiritual—and in their struggle tore [his] soul apart."[40] Augustine's agony in what is undoubtedly his personal Gethsemane is caused by two wills driven towards different ends—his conversion would amount to nothing less than bringing the one in agreement with the other, that is, in his will's self-surrender to God's, in saying, like Christ, "yet not as I will, but as you will" (Mt. 26:39). Had St. Maximus read the *Confessions*, he would have recognized that what is at stake in St. Augustine's struggle is not much different than what he had understood to be the case in the Lord's prayer before his passion, namely, the self-surrendering of his human will to his Divine—so Augustine's conversion constitutes at the same time a unification of his *gnomic* will[41] to his natural will as it was ordered by God and toward God.[42] Yet, please notice, this unification, if it is to be achieved, it can be brought about only by the *gnomic* will.[43] That there are *not* two natures here at fight, an evil one against a good one, is made emphatically clear in *Confessions* VIII.10.22. For St. Augustine, ever conscious of the lurking dangers of Manichaeism, the natural capacity to will, though tainted by Adam's sin, could not be anything else but good. The will's resistance to itself and its subsequent splitting into two opposing wills—this monstrosity (*hoc monstrum*—repeated four times in the space of a short paragraph) is only the result of a will which, bedazzled by its own alleged freedom—the *posse peccare* of the second book's Edenesque garden of stolen pears, moved steadily away from God and, at the same time, away from its own nature.

In the dramatic description of the two wills, St. Augustine succeeded in remaining faithful to his own (as much as the universally human) experience of sin while avoiding the Manichean temptation of justifying evil by positing an evil God. Similarly, I would like to argue, St. Maximus's theory of two wills exculpates human nature, and together with it the natural will, from any Origenist views that would deem what is created as ontologically and morally inferior.[44] (It can be argued that Maximus uses the name of Mani as cryptic allusion to Origen—whom he mentions by name only once).[45]

It is at this point that one can properly evaluate St. Maximus's distinction between a natural and a *gnomic* will. For by positing the *gnomic* will as distinct from natural will, St. Maximus did not divorce human will from human nature nor he allowed the anonymity of nature to engulf the precious particularity that the Gospel had set higher than the abstractions of the Academy, but he struck a balance between the existential experience of evil and the metaphysical goodness of the creation; he affirmed motion's teleological perfection without denying man's freedom from the blindness of necessity; he acknowledged man's fallibility yet he upheld man's ability to receive God's grace and to accept the divine invitation to salvation; he admitted the premises of *apocatastasis* without the obligation to follow its conclusion, for "since nature and person are not identical, the restoration of nature does not of necessity entail the complete restoration of every person."[46]

In the space, the *diastema*, between archeological being and eschatological ever-well-being, we find ourselves in the field where history, our history, is decided. This is the space of time, the time of action as well as of reflection, a reflection that is possible only thanks to humankind's opening to time. However, "man is not just temporal; he *is* Time."[47] To say this is only another way of saying what Augustine had already discovered in the eleventh book of the *Confessions*, namely that time is a property of humanity in its capacity to remember (past), to attend (present) and to expect (future). Those three ecstasies of time, as Heidegger would later call them, form another Augustinian trinity, that of being, knowledge and will. Their parallelism would reveal the correspondence of past with being, of present with knowledge, and of the future with the will. It is on this point that the thought of Augustine and Maximus converges most decisively. For the refutation of Manichaeism (for Augustine) or of Origenism (for Maximus) had made necessary to dispel the Greek cyclical notion of time by offering a corrective in the form of a new conception—rectilinear and dynamic—of history. The difference between these two theories of time is centered on the problem of the will. For a cyclical notion of history is bereft of a true future and without a future the "I will" of the will makes little sense.

Therefore, Maximus's cosmological principle expressed in the triad coming-to-being, motion, and coming-to-rest (*genesis-kinesis-stasis*) finds its application and mirror-image on the personal level as well, where the will is the motion that moves human nature "whose origin is *before* the creature's own being and whose goal is *beyond* it"[48] to that rest which only the alignment of human will with the divine will could provide. Moreover, as in his cosmological view, rest is not simply the cessation of motion but, in fact, its intensification, so with the human will whose willful self-surrender to God's will finds its fulfillment, a fulfillment that will never know any satiety. The connection between the cosmological and the personal view, the connection between man and the cosmos, or better yet, the overlapping symmetry between history (from creation to the eschaton) and the personal history of each one of us is the idea that provides the *Confessions* with its structure and thematic unity. It should suffice to point briefly to the three gardens of the *Confessions* as demarcating the movement from Eden (Book II) to Gethsemane (Book VIII) to the kingdom to come (Book IX). Between the two first gardens, precisely at the middle of the narration that stretches from Book I to Book IX, that is, at the middle of the Book V, Augustine, like the people of Israel, crosses over the waters of the Mediterranean on what is his personal exodus from the land of his captivity to sin (Africa) to the promised land of his salvation through baptism (Italy). Thus, the story of humanity's falling away from God and its return to God is told again in and through the details of Augustine's personal history. When the *Confessions* begin again, in Book XI, by rehearsing the beginnings of Books I and II, the vision that propels Augustine is again that of history, from creation ("from the beginning when you made heaven and earth") to eschatological re-creation ("to that everlasting reign when we shall be with you in your holy city") (XI.2.3, 286). Again this trajectory is understood in terms of his own movement from being to well-being ("you . . . have granted me first to exist, that I may enjoy well-being," XIII.1.1, 342).

The crucial question, however, in my opinion, concerns the eschatological destiny of will—that is, whether the will in general and *gnomic* will in particular, will in some form, however transformed or "glorified," survive at the eschaton? Will or will not the creaturely character of the creature abide eschatologically?[49] That is—does the created/uncreated distinction still hold at the end of times and in God's kingdom? If we want the Chalcedonean formula to express an eternal truth, that is, if we want Christ to be such an eternal truth, then it must. And it must, because without the hypostatic union between the uncreated divine nature and the created human nature Christ is not Christ anymore. As long as there is Christ the creation itself, and thus the created/uncreated distinction, cannot be abolished.[50]

Depending on how one is prepared to answer the questions we posed above, it can be safely predicted what view one would take with regard to the *gnomic* will's destiny: if the created/uncreated distinction will be abolished for the sake of some eschatological panentheism (cf., 1 Cor. 15:28), then *gnōmē* and the person are obstacles in this process of deification and they too must be overcome. If, on the other hand, the distinction somehow still remains, then one cannot dispense with *gnōmē*.

In the Christian East there has always been an emphasis on a strong understanding of deification—but such deification is by grace and not by nature.[51] Yet, does such a strong understanding of the deification of nature mean, in light of the created/uncreated distinction, that human nature, as deified, is eschatologically overcome not only in its fallen, but also in its creaturely character? I would opt for saying that the human nature is perfected, even though its creaturely character remains.

That nature as created *by God* is susceptible to such deification (*capax Dei*) is, of course, without question. Yet, as *created* by God it is also subject to limitations that made impossible for it to reach such a goal by itself—naturally—otherwise we are confronted with the dangers of Pelagianism. As created by God, nature has been given both a direction as well as the impetus that will carry it towards such a goal. It is us, I admit, with our *gnomic* wills that keep taking detours away from that aim, delaying the process of our salvation—yet, I would also say that these long detours and the resistance that they represent on our part against God's plan are somehow instrumental to our salvation and that, therefore, not only our *gnōmē* would not be abolished, but even its history, which is nothing else than history itself, must be upheld and preserved.

God, I assume, does not want to save us away from history—by some Neo-Platonist *epistrophē* to the One—but to save our history, especially because, since the incarnation, it has been also *his* history. If this is, indeed, a fundamental tenet of Christian faith, then God does not come into history in order to save the human nature, but rather in order to save us *from* the anonymity of our nature; a freedom not so much *from* nature itself, but rather a freedom *for* nature to leave behind the shackles of self-love (φιλαυτία) for the sake of relating through love with the other and with God.[52]

It is important to raise the seemingly obvious question of "what is saved?" If God does not save the human person but the human nature as a genus and as an abstraction, then he is not anymore the God "of Abraham, and of Isaac, and of Jacob" but the God of the Platonic Ideas. Another way to say this would be to ask: "what has fallen?"

The human nature or the human person? Both questions, however, operated on an unnecessary and superficial distinction for it is obvious that it is the human nature that is restored by the salvation of the person, as it was the human nature that had fallen by the sin of a man. It is at this point that one can fully appreciate the false dichotomies to which the separation of nature from person and of person from nature would inevitably lead.

Since for St. Maximus the *logos* of nature is the principle of unity, while the will, and especially what he later called *gnomic* will, is an element of differentiation, distinction, and even division, his eschatological vision of a humanity united not only with itself but also with God seems to suggest the eradication of *gnomic* will or, at the very least, its complete appropriation by nature. So, in his *Letter 2*, he speaks of *one* nature and *one* will "with God and with one another"—a goal that will be achieved when love persuades *gnōmē* "to follow nature and not in any way to be at variance with the *logos* of nature."[53] It was passages like this one that prompted Lars Thunberg to observe that "one might get the impression that Maximus was arguing in favour of the idea that the human will should in the end be entirely swallowed up by the divine will."[54] How can we, from within the theological vision of St. Maximus, maintain a balance between unity and difference, between natural communion and personal otherness?

In St. Maximus's theology of two wills a fundamental experience is reflected, namely, the experience of a duality that abides between *what* I am, my nature, and *who* I am, my personhood. Between the *what* (λόγος τῆς φύσεως) and the *who* or the *how* (τρόπος τῆς ὑπάρξεως) one should come to recognize what is most abstract and that which is most concrete. In each one of us two absolutes come together—the universality of nature and the particularity of personhood. Insofar as I am a human being I am no-one and everyone. Everyone, because nature is common, consubstantially common with the humankind, yet no-one, for humanity cannot assume the position implied and necessitated for him who speaks in the first person. On the other hand, as a person I am irreducibly myself—a position uniquely posited and irreplaceable. Between these two absolutes there is a dialectic relation: if I am myself that is only insofar as I am a human being, an instantiation of the human nature; conversely, if there is a human nature, if one can think and speak of such an abstraction, that is only because of particular persons, otherwise "human nature" would have been an empty concept. The idea of a dialectical relation between nature and self might suggest certain symmetry. Yet, the person not only enjoys an ontological precedence over the community of nature, even though, in turn, it is constituted by such a community, it is also always more than its nature; the person is the surplus of an existence that refuses to be completely identified with its essence. Once more, it is useful to repeat here the Kierkegaardian formula that "[a]t every moment, the individual is both himself [*qua* individual] and the race [by virtue of human nature]."[55]

The *distance* of which we get a glimpse through the ontological difference between existence and existents, the distance between myself and my nature, becomes diminished in those moments where I act only naturally—when I am absorbed to the natural mandates of eating, sleeping, etc.[56] In such moments, life becomes effortless or light as in Kundera's *Unbearable Lightness of Being*. Incidentally, but not accidentally, these are also the same moments where one becomes less personal and, one could argue, less human. On the other hand, being oneself becomes invariably a burden that needs to be taken up as one takes up his cross.[57] Being a person—personal being—constitutes and is constituted always by an agony.[58] It is not accidental that St. Maximus's discussion of such a personal will turns always to the exemplar of Christ's agony at the Garden of Gethsemane.[59]

Is humankind meant—either in its alleged prelapsarian Paradise or in the eschatological perfection—to be only a *natural* being, and if so, can we still speak of a human as distinct from the animal? Is not the fall precisely this: the fall from the animal state, or the fall of the animal to humanity, where the sting of consciousness is felt at the moment when "their eyes were opened, and they realized they were naked" (Gen. 3:7)? Is humanity, the very consciousness whose inception is herald by the realization of my nakedness, the result of sin or is sin a possibility only for such a naked and fallen being? Are we to lament or perhaps rejoice over this *felix culpa* that gave humanity to man even before it gave humanity to God? And is not this ability to see myself and my sin, the ability to reflect on my sinfulness the very occasion that sets me on the path to repentance and thus to the ultimate undoing of sin (cf., chapter 5)? Man's ability to *see himself*—the ability to turn his reflective gaze upon himself—as signaled by the verse already quoted from Genesis, points at the distance of myself from myself, the distance between the I and the nature that continuously claims it for herself: that is, to the advent of consciousness. The distance that consciousness *is* appears now as an essential and fundamental human characteristic so much so that one could not do away with it without, at the same time, getting rid of man as such.

Thus, our initial question about the eschatological survival of the creature's creaturely character needs now to be re-posited more specifically with reference to consciousness. Will there be a consciousness at the eschaton, that is, will time as *distentio ipsius animi* continue to characterize our existence in God's kingdom or are we to assume that we shall be "like the angels in heaven" (Mt. 22:30; Mk. 12:25)? We know relevantly little about angels in order to take their existence as a model of our life after the common resurrection.[60] Yet they seem to be a reasonable example insofar as they are, like us, created beings, but unlike us they do not suffer the vicissitudes of time. I appeal to St. Augustine's authority once more, for whom an angel:

> (…) shows no trace of mutability at any point, for it is bound fast by the whole strength of its love to you, who are always present to it; and having nothing to expect in the future, nor any memories to relegate to the past, it is neither affected by change nor a prey to distended consciousness [*nec in tempore ulla distenditur*].[61]

The opinion that humanity eschatologically will be without memory, and therefore without consciousness, it would seem to find John Zizioulas in agreement. In his article "Eschatology and Existence,"[62] he argues that at the eschaton the soul will *not* retain its ability to remember.[63] Is this loss of memory restricted to the memory of sins only or does it concern every memory in general? On the other hand, the eschatological parables of the Gospel (e.g., Lk. 16:19–31) seem to suggest the preservation of memory. The rich man remembered the poor Lazarus and the relatives he had left behind. When Zizioulas writes in the same article that "the dead are not separated from the relations that determined their historical existence, on the contrary, it is those relations that will ultimately judge their eternal future,"[64] he seems to entertain the eschatological existence of memory, otherwise of what good would those relations be if one does not remember them? I would argue that we shall remember even what is now, *sub specie tempore*, perceived as "evil"—but it will be remembered not *as* evil, for through the perspective of time that the eschaton will afford us, indeed through the perspective of the end of times, what was *previously*—namely, at the beginning and at the instant—experienced as evil will be at the end seen with a different understanding. I bring as an example the Lord's passion which, when it happened, was undoubtedly perceived as the ultimate evil. Yet, that same event is now commemorated in the Eucharist, which is the prefiguration of our eschatological understanding, as the source of our salvation. The same event is presented quite differently "at the moment" and through the distance that time affords us. Finally, if the Eucharist is a foreshadowing of the eschaton so much so that what is still future (historically) can be remembered as having taken place already, that is, if the future can be evoked in the present as past, then, are we not allowed to assume that in the eschatological future the past could also be evoked as present?

With or without memory, humanity cannot exist without temporality: the distinction that Zizioulas introduces in that same study on eschatology, a distinction between two eternities and between two times, is very useful.[65] The eternity of man, who as a creature had had a beginning, is not the same as the ageless and motionless eternity of God. The former will be enjoyed by grace, but not at the expense of the creaturely character of the creation. The beginning of creation gave to everything that is created the *permanent* characteristic of motion. "For everything that comes into existence is subject to movement" St. Maximus reminds us.[66] Surely, that movement will eschatologically come to rest in God, yet God's infinity as well as the soul's natural definition by motion allows St. Maximus to speak of a rest that is ever-moving (στάσις ἀεικίνητος).[67] The two understandings of motion that are here implied can be said to correspond to the two distinct experiences of times: one during history that is characterized by distance (διάστημα), the other at the eschaton, *epectatic* but perhaps not *diastematic* (to use these two concepts of St. Gregory of Nyssa). The former separates and divides, the latter re-collects and unites. The perceptive reader would perhaps recognize in the distinction between these two experiences of temporality an analogy to Maximus's theology of two wills. St. Maximus dedicates a large section of his first *Theological Opusculum* (to the priest Marinus) to the eschatological destiny of the will. There he makes quite clear that the *gnomic* will will be retained, for he emphatically maintains that in God's kingdom there will be no identity either of God's will itself with that of the saints, or, in fact, of the will of the saints itself with each other, but only a convergence of what they will be willing:

> Not every human's will will be one [μία] with respect to the mode of its motion [τρόπῳ τῷ κατὰ τὴν κίνησιν]. And at no point will the will of God and that of the saints become one in all its aspects [κατὰ πάντα τρόπον] as it seemed to some, even though the objective of God's will [τὸ θεληθέν] and that of the saints is one, namely the salvation of the elect, that being a divine goal and an end preconceived before

> all ages, and a point about which the will of the saints among themselves and the will of the God who saves them will converge [γενήσεται σύμβασις] ... For God's will desires by its nature the salvation of men, while, on the other hand, humanity wills by nature its salvation, thus, that which saves and that which is saved can never be the same [ταυτὸν], even if the goal of both is everyone's salvation, as proposed by God and chosen by men.[68]

Then, St. Maximus goes on to argue *ad absurdum* what would have been the untenable conclusions that one would be forced to draw if we were to entertain such an identity between the wills of the saints or their will and that of God. It seems that for St. Maximus our two wills will retain their integrity in the age to come as much as in the present age, or better yet, even more then, when the discord between them will have been brought to harmony with each other and immutability (ἀτρεψία) with regards to their object of desire.

In the mature articulation on the subject that *Opusculum 3* represents, St. Maximus differentiates between the two wills by defining natural will only as a possibility, as an ability that is actualized (after an Aristotelian fashion) by a particular some-one, a willing one (ὁ θέλων). In order to elucidate his distinction, St. Maximus provides us with a telling example which, by now, we should come to expect it, namely that of language:

> (...) to be disposed by nature to will and to will are not the same thing, as it is not the same thing to be disposed by nature to speak and to speak. For the capacity for speaking is always naturally there, but one does not always speak, since what belongs to the essence is contained in the principle of the nature, while what belongs to the wish is shaped by the intention [*gnōmē*] of the one who speaks. So being able to speak always belongs to the nature, but *how* you speak belongs to the *hypostasis*. So it is with being disposed by nature to will and willing.[69]

Even if, at the end of times, we all will say the same thing, namely "your will be done," as we all say now the "amen" in the liturgical prefiguration of the *eschaton*, that unison does not obliterate difference, for "how you speak belongs to the hypostasis" and thus we are allowed to utter our prayers and our "amens" not only in our languages and idioms, but also with our distinct accents. Will, like language—to continue borrowing from St. Maximus's example—is highly idiomatic as much as it is hypostatic. No one doubts that the capacity to speak is endowed by nature—yet, paradoxically, a nature that is not enhypostasized in the human person is mute. Neither the "your will be done" of the Lord's Prayer, nor the "*maranatha*" of the Church is or can be uttered by nature. In the great conversion between God and humanity that begun with the world's creation and will continue in the *epektasis* of "the ages of ages" only persons can be partakers.

Nevertheless, the foregoing discussing has afforded us with a glimpse to a problem which, in spite of everything said here, remains: could a finite being be infinitely?

7

The Time of the Body

The self is always posited twice—for otherwise it would have been imperceptible to itself. Even from a merely linguistic level, the "con-" of *con*sciousness already alludes to such duplicity. In the previous chapter, we saw that the duplicity of consciousness is located and occasioned by the will, in particular of a will that sets itself against itself as the experience of sin reveals. We can now say more explicitly that the name of the non-coincidence of the self with itself, of this opening that runs through consciousness and makes it possible while, in turn, maintained by consciousness itself, is desire. Put otherwise, the non-coincidence of the first two gardens, the difference between Thagaste and Milan, Eden and Gethsemane, is desire. Desire as a factor that differentiates, but also defers (in Derrida's neologism, *différance*): it delays my arrival to myself. "The delay deepens into a logically inexplicable differ*a*nce, that of the desire that contradicts itself."[1] Desire, therefore, emerges as a category of temporality insofar as, through latency, it temporalizes the self.

All these themes are announced together in the opening paragraph of the fourth book of the *Confessions*:

> *Throughout those nine years*, from my nineteenth to my twenty-eight year, I and others like me were *seduced and seducers, deceived ourselves and deceivers of others amid a welter of desires*: publicly through the arts reputed "liberal," and secretly under the false name of religion. In the one were arrogant, in the other superstitious, and in both futile (. . .). These ends I pursued, these things I did, in the company of friends who *through me and with me were alike deceived.*
>
> IV.1.1, p. 92

It is necessary to expose the duplicity upon which subjectivity rests by describing its pretenses in order to allow for the primacy of intersubjectivity as diachrony to emerge.[2] This is the task of the present chapter.

I. The theater of self-deception: curiosity, narcissism, and melancholy

Vision presupposes distance and it is rightly regarded as the most philosophical of senses, as *theoria* or contemplation ranks first of the so-called "spiritual senses." Yet,

Augustine in the course of the *Confessions* becomes increasingly suspicious of the *concupiscentia oculorum*, the lust of the eyes, in its double desire to see and to know, especially of what ought not to be seen and of that which cannot be known. Vision, then, translates into a curiosity which is never quite innocent, insofar as it collects under a single concept the figures of both Adam—the serpentine promise was after all "when you eat from it *your eyes will be opened*, and you will be like God, *knowing* good and evil" (Gen. 3:5), of which St. Augustine writes "aperti sunt oculi eorum: audax curiositas mota est," (*Gn. Lit.* 11.31.41)—and Zacchaeus (who, like Adam, "ἐζήτει ἰδεῖν," Lk. 19:3). Adam's curiosity led to that sin which has been called since Augustine "original" and for whom it was always somehow associated with the *concupiscentia carnis*.

Zacchaeus's curiosity, on the other hand, was exemplary of the sin of *superbia vitae*. The Lucan passages suggests so much by specifying Zacchaeus as "chief tax collector" and "rich" but also by the fact that he "ran ahead [προδραμὼν εἰς τὸ ἔμπροσθεν] and climbed up" the sycamore tree in order to see Jesus. There are frequent allusions to Zacchaeus throughout the *Confessions* and always associated with pride, especially pride of knowledge, such as it was most evidently found among the Manicheans (cf., IV.12.19 and V.3.5). In the typology of the *Confessions* Zacchaeus becomes a symbol for the Manichean curiosity (*concupiscentia oculorum*).

Each of these two scriptural examples are also connected to a tree: Adam eats from the tree of the knowledge of good and evil, while Zacchaeus climbs up a sycamore tree. Thus, vision organizes temptation, that is the possibility of sinfulness, in its threefold manifestation as recorded in 1 John 2:16: "For all that is in the world—the lust of the flesh [ἐπιθυμία τῆς σαρκὸς, *concupiscentia carnis*], the lust of the eyes [ἐπιθυμία τῶν ὀφθαλμῶν, *concupiscentia oculorum*], and the pride of life [ἀλλαζονεία τοῦ βίου, *superbia vitae*]—is not of the Father but is of the world." This tripartition of sinfulness provides St. Augustine with one of the most enduring patterns in accordance to which he composes his seminal text.

For example, in the opening passages of Book III, Augustine rehearses the Johannine threefold as *concupiscentia carnis* in 1.1, carrying over the theme of sexuality introduced by Book II; as *concupiscentia oculorum* in 2.2–4 (in the examples of *sacrilege curiositate*, such as the attraction to the theater, etc.); and as *superbia vitae* in 3.5–6 through the examples of *fora litigiosa*, *tyfo*, the "Eversores," etc. From among these three lusts it is that of the eyes which receives the most extensive treatment. The lust of the eyes can manifest itself in different ways:

a) In the desire and enjoyment of watching the suffering of others;
b) In the desire and enjoyment of watching the suffering of oneself;
c) In the desire to know or to see what is off-limits (in which case one sees or knows with a relative impunity for one is not seen back, or, alternatively in order to gain access to a secret one can offer a sacrifice, a small animal or oneself).

Common to all, however, is the attraction to the spectacle and the spectacular, an attraction that is exercised through the mediation of the image. They are constitutive of the imaginary. Thus, the panorama of vision encompasses the whole theater of human sinfulness. And what is in the root of it all is a certain narcissistic element for which, thanks to the distance opened up by vision, the other, even the Other per excellence, is turned into an object of my desire, captured by the idolatrous motivation to see and to possess even if only through a distance. The very distances that constitutes the self's self-defense against love's painful wounds.

> When the ego assumes the features of the object, it is forcing itself, so to speak, upon the id as a love-object and is trying to make good the id's loss by saying: "Look, you can love me too—I am so like the object."[3]

In this process of sublimation, the ego supplements the object of its (i.e., the id's) desire with itself. Yet, the very condition of this desire precludes it from been fulfilled or satisfied, for to traverse or close the distance that separates it from its desired object (in this case, itself) would cancel out that very narcissistic desire. Here, then, we meet the predicament of a desire that feeds on its (empty) self and thus starves itself to death quite unable to die (death instinct). Thus, desire—at least narcissistic desire—perpetuates the bifurcation of the self against its own pretenses. Furthermore, it is quite disturbing that the ego is willing to assume the features of the id's love-object, which is nothing else that those of *das Ding*. More on this, however, in the next chapter.

The curious fact of enjoying one's sadness was first announced in Book I, when Augustine had remarked about that kind of literature that attracted him "I was saddened at being prevented from reading what would sadden me" (I.13.21, 53). Then in Book II we meet the development of this idea as transferred in himself: "I was in love with my own ruin, in love with decay: not with the thing for which I was falling into decay but with decay itself" (II.4.9, 68). Sadness and the love for self-destruction come together in the short mediation on the attractiveness of sad spectacles: "I was held spellbound by theatrical shows full of images that mirrored my own wretched plight and further fueled the fire within me. Why is it that one likes being moved to grief at the sight of sad or tragic events on stage . . .? In the capacity of spectator one welcomes sad feelings; in fact, the sadness itself is the pleasure" (III.2.2, 76). Could Augustine's fascination with theater have been motivated by his search for a narrative that would have re-covered some forgotten trauma? Sometimes vicarious re-living makes possible the experience and the "working-through" of trauma, especially of an impossible trauma, as I suspect St. Augustine's was.

I would like to bring to our discussion the testimony of a classical text that may shed some light on the complex problem of *curiositas*. It is itself a spectacle, a theatrical play like the ones that young Augustine enjoyed, namely Euripides' tragedy *Bacchae*. Dionysus punishment of Pentheus, who refuses to acknowledge the divinity of the new god, consisted precisely in the employment of Pentheus's curiosity to see the maenads. It was this desire that Dionysus uses in order to drag Pentheus to his destruction. Dionysus promises him to lead him to a place where he would be able to see them (810–46). Once they arrive, Pentheus asks to climb on a pine-tree in order to better see the maenads—and perhaps in order to see without being seen. Dionysus bends the tree but once Pentheus is up, Dionysus calls on the maenads, among whom is Pentheus's own mother, who seized him and tore him apart (1048–1147). The scene is suggestive of a sexual desire. Pentheus desires to penetrate, even if only with his gaze, the secret place to which only women have access. Curiosity here is coupled with voyeurism in forming what is essentially pornographic. In order to satisfy his desire, however, Pentheus has to undergo a sex change of sorts by disguising himself into a maenad, that is, into a woman. Thus, he becomes the object of his desire (narcissism). Insofar as, however, the women Pentheus desires to see are not any women, but maenads, that is, devotees of Dionysus, gathered in worship of their god, his curiosity takes up a new sense, that of the profanation of the mysteries, as he remains an uninitiated intruder (sacrilege). Euripides' text sets before us the complexities of the fascination that underlines curiosity. The profanation of the secret/sacred is motivated by a phallic virility—the pine-tree on which Pentheus, like Zacchaeus, ascends—and leads him to destruction.

The imaginary leads us to the question of narcissism. The narcissistic self, Freud tells us, has withdrawn into the sphere of the imaginary.[4] His objects of desire are retained "in phantasy," having "substituted for actual objects imaginary objects found on memories . . ."[5] It is worth citing here Freud's list of narcissistic love. Thus, a person may love narcissistically:

a) What he is himself (actually himself).
b) What he once was.
c) What he would like to be.
d) Someone who was once part of himself.[6]

One could locate with relative ease St. Augustine's youth in both d) with reference to the dead friend of Book IV and c) with reference to Hierius to whom he dedicated his first book, *De Pulchro et de Apto*. In fact, since both episodes are part of the same narrative of Book IV, we could say that Augustine's newly found admiration for the idolized Hierius functions as a substitution that aims at helping him cope with his grief over the lost friend to whom he was attached with such a narcissistic love. The friend was thought precisely as part of his own self:

> I was . . . still more amazed that I could go on living myself when he was dead—I, who had been like another self to him. (. . .) I shrank from life with loathing because I could not bear to be only half alive; and perhaps I was so afraid of death because I did not want the whole of him to die . . .
>
> IV.6.11, 99[7]

On the other hand, Augustine knows Hierius only by reputation ("I did not know him personally, but I had come to esteem him for his splendid reputation . . ." (IV.13.20, 105)). Without the downfalls of intimacy—downfalls that he tasted in the case of the dead friend—Augustine could admire Hierius from the safety of the distance between two strangers. And what admires in him is precisely himself: the kind of person whom he would have liked to be one day:

> All the same, why was the esteem in which I held Hierius not like that evoked, say, by some noble charioteer, or a gladiator made widely famous by popular

> enthusiasm, but something far different, more serious, and akin to the commendation I hoped to win myself?
>
> IV.14.22, 106

Yet, where does his love for tragic plays belong? Where does this desire to suffer (*vult dolere*, III.2.2) come from? What makes the vicarious experience of suffering pleasant? To answer these questions we need to take into our consideration Freud's *Beyond the Pleasure Principle*, a book which, in its discussion of infancy and children's games (pp. 10–25), memory (pp. 26–39), and time (pp. 31–2), offers thematically a close resemblance to the *Confessions*. Freud's analysis is motivated by a similar paradox as that which puzzles the bishop of Hippo in the opening of Book III. For the wounded warriors who returned from the front of the Great War were beset by recurring dreams in which they re-live (as in theater[8]) their original trauma. Freud's answer is quite interesting: by repeating the traumatic experience those dreams sought to exteriorize the internal pain:

> ...a particular way is adopted of dealing with any internal excitations which produce too great an increase of unpleasure: there is a tendency to treat them as though they were acting, not from the inside, but from the outside, so that it may be possible to bring the shield against stimuli into operation as a means of defence against them.[9]

By exteriorizing a pain that was primary felt inside, "[t]hese dreams are endeavouring to master the stimulus retrospectively, by developing the anxiety whose omission was the cause of the traumatic neurosis."[10] Nevertheless, could we apply, all-too-neatly perhaps, this theory on Augustine's fascination with drama? What could have been his traumatic experience? Returning, to Freud's study on narcissism, we find a helpful insight at the end of a series of comparisons that Freud's attempts of the narcissistic self with melancholia and hypochondria (what is common in all three cases is what Freud calls the "introversion of the libido" or a "desexualization"[11] so characteristic of St. Augustine's "holy hypochondria" to use a term aptly coined by Hamann[12]):

> The answer which would follow from our line of thought would once more be that we are so impelled when the cathexis of the ego with libido exceeds a certain degree. A strong egoism is a protection against disease, but in the last resort we must begin to love in order that we may not fall ill, *and must fall ill if*, in consequence of frustration, *we cannot love.*[13]

Augustine's love with love (*amans amare*) seems to suggest a frustration in his desire as a result of which he must "fall ill"—that is, he must simulate pain as an antidote to a mute trauma, or rather, because no trauma was there to justify the condition that prohibited him for being genuinely in love, then, all the more, a trauma had to be invented, provoked, and, at last resort, provided by his imagination.[14]

Reviewing the foregoing discussion, we seem to reach the following conclusion: *superbia vitae* (pride) is the sublimated form of *concupiscentia oculorum* and *concupiscentia oculorum* is the result of a frustrated *concupiscentia carnis*. What the subtler and loftier temptations seek in their respective spiritualized objects is the satisfaction that has failed to satisfy the flesh. Yet, we could not fail to notice that the more spiritualized a desire is the more serious or "sinful" it becomes. The spiritualization of sin is comparable to the "spiritualization" of crime—an example that could serve to elucidate our point: what sets the detective apart from the crime that he investigates is only the criminal act, but not the crime itself. For the detective, especially if he is to be effective in resolving the crime, is as much obsessed with the crime as the criminal, if not more: more because he has no *motive*, like Augustine's theft lacked a motive, other than the satisfaction of his curiosity and the self-exaltation of his intellectual abilities. The detective "commits" the crime by solving it, simply because he can. The detective commits the criminal act, albeit only from the distance and safety afforded by his mind. The crime, if it is to be resolved, must take place again in all its gruesome details, step by step, but this time in the detective's mind. If the detective's re-enactment of the crime *in mente* seems innocent, this is only in the eyes of the law for which actions and not thoughts matter. Sin, however, insofar as it is not restricted to the jurisdiction of the judicial alone, is not exculpated by becoming spiritualized—in fact, in its spiritualization sin becomes more insidious by becoming internalized. We are reminded here again of one of the archetypal murder mysteries, that of *Oedipus Rex*: its uniqueness consisting in the fact that the detective who so willingly undertakes the investigation of Laius's murder is the murderer himself. Perhaps Sophocles' intention was to remind us that every persecutor is already implicated in the crime he investigates. Or, to revisit, Father Zosima's worlds: it is us "who are guilty on behalf of all and for all."[15]

Here too the incarnation of the spiritual suggests itself as the recommended course of healing. Spiritualizing or sublimating does not offer a re-solution, but instead, a complication or, as we could say, a *complexification*. The resolution is to be found in and be effected by a descent into the bodily. This descent to the bodily amounts, as I would like to claim, to the therapeutic realization of the hidden complicity between the super-ego and the id, between the higher and the lower, as it is with all hierarchical psychological tripartitions, be they Platonic or Freudian.

> What has belonged to the lowest part of the mental life of each of us is changed, through the formation of the ideal, into what is highest in the human mind by our scale of values.[16]
>
> The abundant communication between the ideal and these Ucs. Instinctual impulses solves the puzzle of how it is that the ideal itself can to a great extent remain unconscious and inaccessible to the ego. The struggle which once raged in the deepest strata of the mind, and was not brought to an end by rapid sublimation and identification, is now continued in a higher region, like the Battle of the Huns in Kaulbach's painting.[17]

Indeed, we are here reminded of Plato's trichotomy of the soul, in line with the soul's image, namely the ideal city of the *Republic*. Yet, the city's stratification rests on a secret; a "noble lie" as Plato himself calls it (414c), for there is no essential difference between the three classes and, therefore, one is left to surmise that, similarly (after all, it is of such similarities than Plato's work consists) the threefold division of the soul is equally arbitrary, or rather worse, that the rational part of the soul is one and the same with its libidinal pole.[18] Furthermore, this descent to the bodily takes a little more concrete form if one were to follow Plato's attempt to locate the different parts of the soul in the anatomy of the human body (see, for example, *Timaeus*, 69e–73a), beginning with the loftier and more rational part that is assigned to the head—remember that the whole of the *Republic* takes part in Cephalus's house whose name simply means "the head"—to the appetitive part that is, appropriately perhaps, located on the antipodes of the head, that is, in the regions below the belly.

St. Augustine's narrative suggests such a descent into the bodily in two main cases: a) by the reversal of the traditional hierarchy of the senses, whereby he places vision at the lowest end and gives priority to touch, as vision's very opposite, insofar as tactile intimacy necessitates the annihilation of distance on which vision operated and b) by the consistent and constant employment of the parable of the prodigal son that would allow him to place a newly discovered emphasis on the flesh.

II. The Time of the Body: Hunger, Eating, Sex

We have seen in the previous chapter that in the *Confessions* a double process of unification—what we called a *universion*—is set in motion: on the one hand, the story of the self's conversion, returning to one God and, at the same time, a "return to oneself" (cf., Lk. 15:17); on the other hand, this narrative of the self's unification itself rests upon the presupposition of a self unified. If the narrative of the *Confessions* is made possible, this is only to the extent that the narrator has collected himself into one. Recollection is essentially the result of such self-collection. What prompts or enables this (re-) collection of the self? As the self's distance from itself is not spatial, the answer lies again with time. Thus, Augustine writes:

> Not with our feet or by traversing great distances do we journey away from you or find our way back. That younger son of yours in the gospel did not hire horses or carriages, nor did he board ships, nor take wing in any visible sense nor put one foot before the other when he journeyed to that far country where he could squander at will the wealth you, his gentle father, had given him at his departure.
> I.18.28, 58

Yet, not by the turning of the mind either, unless it is a mind affected by the *after* (*meta-noia*). We recall again that monstrosity which Augustine so aptly describes in Book VIII: "The mind commands the body and is instantly obeyed; the mind commands itself, and meets with resistance" (VIII, 9.21, 201). Can we reverse this chain of command? Can we say that the body could perhaps command the mind and that mind would obey? The prodigal's *metanoia* that set him upon the journey back to the paternal house was not a mental, spiritual, or psychological affectation, but physical, as physical and visceral as hunger can be.[19] The task in the *Confessions* is not to spiritualize the flesh, but to incarnate the spirit and in this regard the verticality of the Greek schema

(Platonic and Neo-Platonic) of ascents and descents has to be abandoned for the sake of a communion between exteriority and interiority. Yet, as long as hunger is only "interior," that is, spiritual, is not even perceived as hunger at all. "I was inwardly starved of that food which is yourself, O my God. Yet this inner famine created no pangs of hunger in me" (III.1.1, 75). Spiritual hunger remains ineffective.

Let us return to the Lucan parable:

> "After he had spent everything, there was a severe famine in that whole country, and he began to be in need [ὑστερεῖσθαι/egere]. So he went and hired himself out to a citizen of that country, who sent him to his fields to feed pigs. He longed to fill his stomach with the pods that the pigs were eating, but no one gave him anything."
>
> Lk. 15:14–16

Notice that from all the images that the Gospel could use in order to convey the state of human misery, the evangelist chose that of *hunger*. Eating is not only a way for recognizing our dependency to each other and to the world—so much for the prodigal son's claim to independence!—but by eating we assimilate the world to ourselves, we turn that which is outside inside. Think of this passage from outside to inside and you will discover that this opposition is nothing else than the exemplification of distance and fragmentation. Ultimately, distance comes down to this opposition between an inside (that I identify with myself) and everything else that is outside me. In eating, however, this wall of separation collapses—when I am hungry I am really hungry for the Other (following Sartre and Levinas)—and eating is one of the ways we have in overcoming our isolation that is the result of being scattered beings. Eating declares—willingly or not, and contra to all our illusionary attempts to self-mastery and independence—my dependence on the world, on the cows which provide me with their meat, but also on the grass that fed the cows, on the water that fertilized the soil on which that grass grew, and so on. "But eating, by contrast, is peaceful and simple; it fully realizes its sincere intention: 'The man who is eating is the most just of men.'"[20] Levinas in *Existence and Existents* juxtaposes a referential totality of food to Heidegger's referential totality of tools: the world of enjoyment takes precedence over the world of action. A referential totality is presented in every meal whereby the entire world is eaten. *When I eat, I eat the world.* But even more than the world, I eat the labor and the effort, the care and the artistry of the people who cultivated, prepared, and cooked my food. No meal is ever solitary—even if I eat alone in the seclusion of my room—every meal is a public and communal event. A community established and referred to by every bite.

It is also a way that can take place only by means of and thanks to our bodies. Obviously, only an embodied being can be hungry and only an embodied being can eat. Contrary to what one might believe, our best chance to overcome the fragmentation that human nature imposes on us is through our bodies. It is our body that abridges the distance that keeps separating us from others, but it is our body that allows us to be united with God—it is not accidental that we *eat* the Eucharist and that the liturgy takes the form of a meal, such as the meal that the Father offers in celebration for the return of the lost son: "Bring the fattened calf and kill it. Let's have a feast and celebrate" (Lk. 15:23). As it is not an accident that we have these gifts on account of the body that Christ took upon himself in his kenotic effort to traverse the land of distance that separated us from him. In Christ God is not any more a god who sees (*theos*), but a god who touches and can be touched even to the folly of being eaten. In eating, I make the world my flesh and my blood, *not* my spirit, but I incarnate or rather I incorporate—the original and literal meaning of incorporation—the world which, otherwise, would have remained an abstraction in spite of its cows, its grass, and its soil. Spiritualization posits a real danger that seeks to cancel out both the incarnation as well the transubstantiation of the Eucharist. Every time one partakes of communion he or she does not change the host into their body but it changes them—in an inverse digestion indicative and exemplary of inverse intentionality—into Christ's body. As Augustine was told: "I am the food of the mature; grow then, and you will eat me. You will not change me into yourself like bodily food: you will be changed into me" (VII.10.16, 173). The Church is turning the world into Christ a mass at a time. There is not room here for the "spirituality" of Gnosis. Without a physical body, our experience of the distance that separates us from each other becomes permanent. It is in this sense that we read in the parable of the rich man and the poor Lazarus that: "between us and you a great chasm has been set in place, so that those who want to go from here to you cannot, nor can anyone cross over from there to us" (Lk. 16:26).

This explains, I believe, the systematic and, at times, elaborate employment of the language of hungering, eating, and feasting in the *Confessions* that begins with Book I and extends with Book X, where memory is compared to the stomach of the mind (X.14.21). It also sheds some light to a broader operation of incarnation set in motion in the *Confessions*, best exemplified perhaps by a reversal in the hierarchy of senses, where touch ranks higher than vision (see, for example, X.6.8, X.8.13, X.9.16, X.12.19, X.21.30) and by the assignment of the same key adjective, "restless" (*inquietum*/*inquieta*) in describing both the spiritual longing of the heart (I.1.1) but also the physical manifestation of his sexual desire in his youthful erections (II.3.6). The desire for God is not independent from the desire for the other human; nor is the desire for rest promised in God's kingdom contrary to the satisfaction that the body seeks. One who has not felt the latter rarely and with difficulty would seek the former.

Finally, hunger entails a temporal element. Despite of satisfying my hunger, in the course of time, I will become hungry *again*. Hunger and its satisfaction by eating demarcates *the before and after of the body's time*. Before repentance's after-thought (*meta-noia*), there is the after-body (*meta-somoia*): a body that becomes hungry is a body affected by time. The proof is again in the Scriptural parable of the prodigal son:

> "After he had spent everything, there was a severe famine in that whole country, and he began to be in need [ὑστερεῖσθαι]."
>
> Lk. 15:14

The Greek verb employed here to denote a need that is most physical, namely hunger, means primarily "to come late," "to be too late." Ὑστερέω implies a lack or want first and foremost in terms of lateness (ὕστερος).[21] The prodigal son was, like Augustine, late ("Late have I love you ..." (X.27.38)). This ethical lateness is first inscribed and awakened in the body.

Allow me to use another example, this time from the world of the Opera. In *La Traviata*'s last act (Act III), Violetta, after reading Giorgio Germont's letter (*Teneste la promessa*) exclaims: *è tardi*! There is an irony in a future that is late: a letter long awaited, a forgiveness never hoped for—finally they arrive but it is too late. It is always already late, as long as there is time or, rather, as soon as there is time, it is too late. For without this lateness, that is without the concept of "being-late" one could not possibly have time as such. Of course, Violetta's "*è tardi*" has a different weight than the before and after that flank any perception of the now. Yet, one is reminded of Kafka's observation that the Messiah will come after he is expected—that is, he will come late, indeed when it is too late, when one does not need him and does not know what to do with him.

> The Messiah will come only when he is no longer necessary, he will come only one day after his arrival, he will not come on the last day, but on the last day of all.[22]

There is a lot one could perhaps say on this notion of a late Messiah. It prompts me to ask at what time in history's span the coming of the Messiah would not have been "too late"? As soon as there is history, it is already too late for salvation. What does it mean to arrive late? Are we not all of us late-comers with respect to our lives, to our present? Isn't it this precisely what moves consciousness, what drives it, the always already failed attempt to catch up—but with what? With itself. Consciousness wants to catch up with itself, to become synchronous with itself even though such synchronicity would signal its death.

We have seen in the last chapter language's double function in the *Confessions* as both the symptom of separation and as the effort to cope with separation by allowing the subject to enter the world of the symbolic substitution—a world shared with other subjects whereby the communication of symbols creates a community in place of a lost unity as well as it allows the subject to communicate its desire to itself and to others (thus desire becomes such only to the extent that becomes articulated). We are reminded at this point of Augustine's frustration as an infant—that is, as a subject that has not yet arrived to language: "I was frustrated in this, because my desires were inside me, while other people were outside and could by no effort of understanding enter my mind" (quia illae [voluntates] *intus* erat, *foris* autem illi, I.6.8, p. 44). The first experience of fallenness (trauma) is the separation from others, in terms of a distance in space.[23] Language here, as in Freud's celebrated example of the child's game while uttering alternatively the words "fort/da," becomes the means to cope with and remedy a painful absence. Thus, the very birth of language is organized by the spatial and temporal interplay between presence and absence—the need to make present, if only in symbolic form, what is absent.[14]

Language's ambiguous state is centered on the supplementary function of signification, where the absent thing is rendered present by its corresponding sign. In doing so, language creates an illusion of presence under which the absence of what is absent remains hidden. Phenomenology's original contribution, however, consisted precisely in the recovery of the absence of the absent, by refusing to substitute it with a quasi-presence (e.g., the linguistic sign, the image "in" the mind, etc.) and by drawing attention to the role that absences play in the intentional life of the consciousness, where every aspect is always given together with a prospect and a retrospect, every presence is surrounded by absences. In short, if consciousness is not limited to the presence of the present—the synchronicity which we have just called consciousness's death—if it is able to intend what is absent, that is by virtue of an absence or an absence-ing. Thus, we can say, together with Ricoeur, that what is language for the psychoanalyst is intentionality for the phenomenologist.[25]

The similarity extends further. As in every act of speech there is always an unspoken remainder, so every act of intentionality remains fundamentally partial and incomplete. Its partiality is twofold and has to do with the mutual embodiment of the thing given and of the intending consciousness. For, as we have seen, by no act of perception I could ever be able to see a thing in all its aspects and from all its angles; and, similarly, there could have been no consciousness that is not embodied and thus, by its very embodiment, limited.

The incompleteness of intentionality is so crucial that in Husserl's writings it becomes the criterion of distinguishing between the unreflective and reflective consciousness:

> . . . a veritable abyss yawns between consciousness and reality. Here, an adumbrated being, not capable of ever becoming given absolutely, merely accidental and relative; there, a necessary and absolute being, essentially incapable of becoming given by virtue of adumbration and appearance.[26]

We could say now without further delay that what makes the difference between reflective and unreflective consciousness, between the consciousness of itself, and the consciousness of the world and its manifestations, is nothing else than the *body*. Yet, we would be equally right in saying that what makes the difference between reflective and unreflective consciousness is nothing else than *time*. For if the presentation of phenomena is necessarily adumbrated, this is because they cannot give themselves all *at once*, but rather only through the unfolding of time—a time that is ultimately grounded in the consciousness itself.[27] How are we to think of this homology between the bodily and the temporal with which phenomenology presents us? We spoke above about the body's time but it would not be accurate to speak of the body's own time, rather time is the body's, even before it is the mind's (internal time) or the world's (objective time). And it is, in fact, through that time that the mind's affliction of self-deception and self-division can be overcome.

We shall leave aside for now Merleau-Ponty's insight that the body in its ambiguous state (neither consciousness nor a thing of the world) is phenomenology's closest approximation to the Freudian unconscious.[28] It is the body, Ricoeur notes, in its capacity for sex that enable us to exist "with no distance between us and ourself, in an experience of completeness exactly contrary to the incompleteness of perception and spoken communication."[29]

Why bring the body and time together? Rather, why do we suggest that the body is the locus, the site of time? The answer lies in the phenomenological observation of imperfection: the very imperfection which Kierkegaard connected with longing—with desire that moves and is moving.[30] There is expectation, and time in general because thanks to embodiment—both ours and the world's—phenomena could never present themselves, give themselves *all at once*. We understand now that this "all at once" (*Zugleich*[31]) would have been nothing less than the ruination of imperfection, it is after all the original demand for perfection (see, for example, the story of the primordial Fall: fallen imperfection is the result of our failure to accept our original imperfection, that is, the very fact of having an origin, a beginning, while, looking toward the future, fallen imperfection is the result of our impatient desire to haste the coming of eschatological perfection before its time, or rather, the desire to do away with the burden of waiting—the desire to "wipe out the duration of time and, with this, any irritating delays."[32] The body puts a halt on those demands whether their orientation is toward the past, in the form of the protological refusal of origin, or toward the future, as the desire for eschatological immediacy and impatience.

Thus, time now can come about—finds its horizon in the space opened up by the body, the place that is precisely embodied in the body. So much is suggested by the story of Genesis where the bodily (the garments of skin) and time as history emerge together. It is also this connection that situates the temporal in the bodily that makes us believe that time will have no end. This does not mean that perfection will be ultimately (even eschatologically) unattainable, but rather that the pleasure of perfection will remain insatiable (*epektasis* contra the Origenist *koros*). Otherwise, the possibility of an eschatological boredom—with all its devastating implications—would have been inescapable.

Andy Warhol's observation that "sex is nostalgia for sex"[33] betrays the connection between temporality and embodiment insofar as nostalgia is the desire to return, to go back to an idealized past that never was, to one's own Ithaca (as another Odysseus), to one's childhood and, ultimately, to the mother's womb. It is, therefore, our desperate effort to escape the burden of existence, to refuse the presence of the present and to cancel out the promise of the future. "Sex is nostalgia." Nostalgia for what? For sex, for the nostalgic itself, the desire to desire a past outside of time, "a past that has no date and can have none."[34]

An act of embodiment as having sex becomes the moment of truth—"from the start the question of pleasure is the question of truth"[35]—insofar as the self-deception enabled precisely by the opening between the reflected and the unreflected, the saying and the said, is eliminated by the closing of that very distance of self from itself by the descent to the bodily. Nevertheless, this descent remains always as something desired but never completely achieved—the impossibility of desire, the impossibility of "existing as body, and nothing but body"[36] is due to the inescapable fact that desire itself is always articulated, that is, desire is fundamentally a demand on the Other, for it is never simply the desire *for* the Other but the desire for the Other's desire, the demand "Desire me!" Only for him who becomes fully and entirely flesh (John, 1:14) and who is one with the utterance "this is my body"—only for him who is this body and nothing but his body, the descent to the bodily has not only become possible but actual in a desire that is not any longer desired but lived. Indeed, the phrase "this is my body" must be understood (to the extent that it can be understood) only as an erotic declaration, in the same way one understands (to the extent that one understands) the phrase "I love you."[37] "This is my body" makes little or no sense if one were to approach it as an informative proposition like "Paris is the capital of France."[38] Rather, the invitation to "take, eat, this is my body" does not inform but it performs: it offers unconditionally and unreservedly the Other's body at a moment—and it could only be for a moment—where the torturous stretching of the soul on the rack of time is closed over a body infinitely condensed because it condenses infinity in a body. Thus, for a moment, *Psyche ist ausgedehnt*—not any more.

The evidence provided by the mystery of the incarnation—understood in the broadest possible sense—regarding the distance that splits our consciousness and which is overcome at the moment of orgasm or prayer can only be momentary. The narrowness of consciousness necessitates our inability to abide in this closure and the inevitability of our return to the realm of history organized by time-difference, that is, by the perpetual deferral of desire. In the order of the symbolic, there can be no such absolute synchronicity as promised by the descent to the bodily. As long as desire remains articulated—and it can never escape such articulation[39]—the difference between the saying and the said remains.

Here, again, we can posit the same question that we asked at the end of our third chapter with respect to St. Augustine's *Confessions*, only now with a better view to a possible answer. That is, why St. Augustine continue to write after Book X, after the end of his history but also after the end of history itself? Or, put otherwise, why history does not end with Christ's resurrection and ascension, but it continues to drag until today? Which is still to say: why is not the celebration of the Eucharist once enough? How could one repeat what seems to be a singular event? Continuity, repetition, time: these questions place before us with urgency the question of time. Only now, thanks to the foregoing discussion, we know that the question of time is inexorably linked to the question of desire and, therefore, to the question of ethics.

* * *

On the basis of the discussion undergone in these two last chapters (chapters 6 and 7) we have come to recognize a certain structural and functional homology between language, desire, and temporality. Among the characteristics of this similarity, we could list their ambiguity in the role they play with respect to the community or communication with others: each one of them—language, most evidently, but also desire and time—functions as movements toward the Other, as foundational conditions of intersubjective relation. At the same time, however, they highlight the difference, or better yet, the distance between the Other and myself. A similar interplay of distance and the movement beyond distance for the sake of some unification is to be observed with respect to the division that runs within the self itself and which has been the subject matter of our discussion in these chapters. Here too, language, desire and, above all, time promise us the recovery that we are forced to perceive as a lost unity, even though one could hardly entertain a time that the self, namely ourselves, ever enjoyed such a unity. Yet, it is again language, desire, and time that effect or, at least, betray the self's duplicity. It is, indeed, puzzling to notice how much each of these factors bears on the other and what a close cluster they form together.

Perhaps the most astonishing parallel among language, desire, and time is the fact that each of them is grounded in a reality that is so different from them (and so similar to itself) that would seem to contradict them. What I mean by this is that one could legitimately trace the origin of language to a reality that itself remains ineffable; the origin of desire—that "object" which propels and attracts the Protean transformations of desire—is itself a thing (*das Ding*) without attributes and a limit to desire—crossing over it would only mean that one, like Oedipus, "desires to know the last word on desire."[40] Finally, the source of time is, if one were to follow Husserl, something atemporal, a reality exempt from the ubiquity of time.

To say that what these observations seek to underline is the old argument (already rehearsed in Plato's *Phaedo*) that contraries come from contraries is to oversimplify it. For what generates language, desire, and time is not merely something contrary to them but, more properly, a negation. What kind of negation this might be, that is, whether it is a negation of lack or of excess, must remain for the time being an open question.

As we notice a homology among language, desire, and time, so we are surprised to see that the realities which, for each of them, relates to them the way a cause is related to its effects, borrows characteristics of the other, as if all three sprung from the same source. There is no need for me to repeat on reverse everything I have already said about the similarities among language, desire, and time in order to make my case. It would suffice to give only one example, which, as it happens, is also a testimonial that further corroborates my position. Freud writing about the id notices a curious characteristic, namely its immunity to the effects of time. He writes:

> The logical laws of thought do not apply in the id, and this is true above all of the law of contradiction. Contrary impulses exist side by side, without cancelling each other out or diminishing each other: at the most they may converge to form compromises under the dominating economic pressure towards the discharge of energy. There is nothing in the id that could be compared with negation; and we perceive with surprise an exception to the philosophical theorem that space and time are necessary forms of our mental acts. There is nothing in the id that corresponds to the idea of time; there is no recognition of the passage of time, and—a thing that is most remarkable and awaits consideration in philosophical thought—no alteration in its mental processes is produced by the passage of time. Wishing impulses which have never passed beyond the id, but impressions, too, which have been sunk into the id by repression, are virtually immortal; after the passage of decades they behave as though they had just occurred.[41]

I think it would not be difficult, on the basis of Freud's remarks, to add to this the id's resistance to articulation (i.e., language).

These observations, on the one hand, legitimize our decision to bring together disciplines that might be taken normally as distinct and unrelated as phenomenology, theology, and psychoanalysis; on the other hand, they force us to ask questions that otherwise would have remained outside the scope of our inquiry. For example, is there a correspondence between the phenomenological understanding of the body and the id? (We mentioned above in passing Merleau-Ponty's insight that indeed this is the case.) Is there a relation between the body and the "absolute flow" from which all temporality in the life of consciousness originates? And, is there a relation between the "absolute flow" and the "ever-moving repose" of Christian eschatology?

8

The Ethics of Desire

I. The return of the Manichean hypothesis: Levinas's radical alterity and Lacan's ethics of desire

At the beginning of Section IV of his *Totality and Infinity* (that is, the section that bears the interesting title "beyond the face"), Levinas brings up the theme of death. Death appears insofar as it constitutes violence, a violence, moreover, that would render truth impossible since the existent—the ontological locus of truth—would be no more. "Will the violence death introduces into this being make truth impossible?" Levinas asks. "Does not the violence of death reduce to silence the subjectivity without which truth could neither be said nor be . . . produced?"[1]

As soon as the violence of death has been raised in the pages of *Totality and Infinity*, it is dismissed for the sake of "love and fecundity" that constitute a "plane where the I bears itself *beyond death*."[2] We are all more or less familiar with the analysis of fecundity that follows. I would like, however, to stop for a moment at this critical juncture where the specter of death appears only to point to its beyond. I am not interested in reading Levinas's movement "beyond death" in opposition to Heidegger's analysis of being-towards-death, although such an opposition is certainly invited. I am rather interested in the strange rapport that is established thereby between death and love—a love that survives death but, as I would like to argue, at the expense of the Other's alterity. I see, therefore, two problems: first, Levinas's discussion of desire is unable to escape Socratic erotics and in particularly what is most dangerous in them (*Symposium*) and secondly, the betrayal of his Judaic tradition which had seen in procreation the mark of death (Genesis). I, finally, believe that the analysis of desire in *Totality and Infinity* contradicts Levinas's earlier project, already underway as early as his *Existence and Existents*, namely the effort to distinguish hypostatic being from the anonymous *il y a*.

Allow me to sketch very briefly the contours of that distinction. "One possesses existence, but is also possessed by it"[3]—in a number of places Levinas refers, quite naively[4] in my opinion, to the mastery that the existent exercises over existence or being, hence the idea of one possessing (one's) being. Yet, being possesses us back. In fact, being's possession of what it is (of the existents) is prior to that of my possession of being. This is, therefore, Levinas's insight: the distinction between the personal "I am" and the impersonal, anonymous "there is" (*il y a*). The two cannot fully overlap and coincide: rather we get a glimpse of the former only in its struggle over against the latter, or, conversely, we get a glimpse of the latter only as the threatening possibility that lurks there for the former. The "I am" is *hypostatic*: this is the term that Levinas chooses. It is a significant term with a long history and a peculiar one. Not only the choice of "hypostasis" differentiates Levinas's metaphysics from Heidegger's *ecstatic* ontology, but it also puts the former in line with a certain tradition that first introduced that distinction, the distinction between *hypostasis* (person) and *physis* (nature).

"Nature" here has a specific meaning that has already become enlightened by Levinas's proceeding discussion. It will be useful, nevertheless, to clarify that "nature" does not refer to what we might call so in everyday language, as we do when referring to forests, mountains, oceans, etc. "All these beings," Levinas writes "are already personal, for they are existents; they already presuppose the category of substantive under which they fit."[5] Rather "nature" means being in its most indeterminable and indefinable state (what "nothing" is for Heidegger). And indeed it is no-thing and above all not a natural thing, for nature itself is, must be, un-natural and uncanny (Freud's *das Ungeheuer*, Plato's *to daimonion*).[6]

This uncanny, demoniac aspect of abstract being is revealed through *horror*, "the horror of Being."[7] For Levinas, horror is prior to anxiety ("being-towards-death") or nausea, since the last two presupposed the existence of a hypostasis, that is, they concern the existent as existent. In horror, however, the existent finds itself confronted with a terror greater than death (cf. Levinas's cryptic observation at the end of Chapter I about "[s]omething that death cannot resolve"[8]). The *there is* in its elemental form is "the impossibility of death, the universality of existence even in its annihilation ... the fatality of irremissible being."[9] The correlate to this horror is, in the language of psychoanalysis, the *death drive*. Following Lacan's idiosyncratic reading of Freud, we could say that the death drive is not, as it is often misunderstood, a death-wish, but rather the impossibility of putting being to death.

An example, through which the death drive manifests itself, could be what we call a *partial object*: a part of a whole that becomes autonomous, assumes a life of its own and subsequently "possesses" the whole (e.g., the hand, the voice). It is strange to think of nature as a partial object—rather it is the very force that drives the undying death of a partial object. In front of nature, understood in this sense, one is powerless. Nothing, not even death, can stop it or put it away. Murder is of no avail (see Levinas's example of Macbeth[10]). Even if you kill this or that, even if you destroy this or that, the natural itself is left unharmed, if not strengthened by my actions.[11] One could offer an example of what we mean by being's "undying death drive" through a story: imagine that the doorbell in your house starts ringing; you go to the door only in order to realize that no one is there. However, the bell continues to ring. You suspect a technical problem (a short-circuit perhaps) and therefore you go to the box that controls the doorbell system. You remove it, and as the frantic ringing goes on, you cut all the wires, you pull the mechanism out. The bell, though, continues to ring.

"The difficulty of separating Being from beings and the tendency to envisage the one in the other are not accidental. They are due to the habit of situating the instant, the atom of time, outside of any event."[12] Here already one finds time. Time as the horizon of being? Rather time as that category which simultaneously enables and obfuscates thinking after being. That habit is responsible for the confusion between Being and beings. What does it mean to "situate the instant ... outside of any event"? It means to think time not as history but as eternity. We read a few lines further "Beginning is already this possession and this activity of being. An instant is not one lump; it is articulated. This articulation is what distinguishes it from the eternal, which is simple and foreign to events."[13] So it is the eternal as "foreign to events" with which one confuses the instant. What does it mean then to view Being under the prism of the eternal instead of the instant? It simply means to be blind to the event-fullness of Being, that is, to be-ing as an event, the event of a beginning. So Levinas writes: "Beginning, origin and birth present a dialectic in which this event in the heart of an instant becomes visible."[14] The instant and the hypostatic share not only the same lexical structure (*in+stare*, to stand in; *hypo+stasis*, to stand under) but also an equivalence in philosophical meaning.

What is at stake here is the contrast between an archaic ontology, one could say pagan as well as static, that envisages being as eternal, outside time and history and, on the other hand, a dynamic ontology, informed by the Judeo-Christian vision of being, which takes the beginning as it point of departure. To these two distinct ontologies correspond a pair of similarly distinct ethics: an ethics that posits good and evil as two contemporaneous principles (which is to say, as two principles indifferent to time), and an ethics that recognizes evil as a moment in the temporalized unfolding of the good. For Greek thought, so the implicit accusation goes, the beginning as an event is unthinkable, for there is always something to be presupposed. Therefore it is a certain philosophy that has "the habit of situating the instant . . . outside of any event," that is, of being unable to think the instant as such (in-stance).

From the beginning of the world to its end: ". . . these are things," Levinas writes in the opening paragraph of chapter II of *Existence and Existents*, "we run up against in the twilight of a world, things which reawaken the ancient obsession with an end of the world."[15] And again: "In the situation of an end of the world the primary relationship which binds us to Being becomes palpable."[16] Being can be thought only from the point of view of a beginning, such a point of view is, however, made impossible insofar as one has already begun (in a sense I was not there at the moment of my birth, let alone at the moment of the world's beginning). Therefore, the beginning can be thought only through "the end of the world." That is our only alternative. For "the end of the world" one does not have, in fact, to wait until the annihilation of the planet or for some similar scientific or apocalyptic scenario. The "end of the world" happens, so to speak, in the everyday, when one is removed from the sum total of relationships with the world that constitute what we call "the world." "For where the continual play of our relations with the world is interrupted we find neither death nor the 'pure ego', but the anonymous state of being."[17]

To think the beginning, therefore, amounts to think of the end. The two are given together in thought. Said differently, to think of the beginning means to think, no matter how opaquely, that out of which what the beginning begins comes from: in the case of the world, nothing; in the case of an existent, being in general. If there is a beginning, there is a moment, an instant. The instant is not a unit in a series that would denote duration, but singular, without precedent and without antecedent. Thus, an instant marks a break, a separation or rather, the upsurge of the new. The new is that which was not before, what has no past. The new owes its novelty to the incisive character of the instant, the momentous of the moment. And this is the beginning.

> One is first struck by the paradoxical nature of this relationship. What begins to be does not exist before having begun, and yet it is what does not exist that must through its beginning give birth to itself, come to itself, without coming from anywhere. Such is the paradoxical character of beginning which is constitutive of an instant. And this should be emphasized. A beginning does not start out of the instant that [proceeds] the beginning; its point of departure is contained in its point of arrival . . .[18]
>
> For the production of an instant of time cannot come from an infinite series, which it would have to traverse, but shows an indifference to that series; it can cut the Gordean knot of time without untying it. It can be, out of itself. That way for an instant to be is to be present. The present is an ignorance of history. In it the infinity of time or of eternity is interrupted and starts up again.[19]

But what begins in the beginning and thanks to it, now it should be clear, hovers over nothingness (or being in general, anonymous being). This will explain two points made later on in Levinas's essay, namely that a) "All emotion is fundamentally vertigo, that vertigo one feels insinuating itself, that *finding oneself over a void*";[20] and b) the reference to the "theory of continuous creation in Descartes and Malebranche," according to which creation is unable "to preserve itself in existence, in its need to resort to divine efficacity at each instant."[21] Levinas goes on to speak of "the drama inherent in an instant itself, its struggle for existence."

Indolence rehearses the beginning. "Indolence is essentially tied up with the beginning [of an action]."[22] And again: "Indolence is an impossibility of beginning, or, if one prefers, it is the effecting of beginning."[23] In a sense, the well-known excuse that sums up the tragic character of the facticity of human existence that we came to an existence not chosen by us is dissipated in the face of indolence, for the decision to exist is made every time one gets off the bed, every time that the burden of existence is willfully taken up.

It is in this revelation of existence "as a burden to be taken up"[24] that "the evil in Being"[25] consists of. But is it calling Being "evil" going a bit far? Indolence and fatigue might indeed reveal the "burden" that being oneself is, since making the decision to exist is required every time and being oneself is never really effortless. But is that enough to justify the characterization of Being as evil? By being ourselves we escape the evil of Being, we recoil from its anonymity—the *there is*. To consider evil a defect of being, or an absence of being is to exculpate Being. "Is there some sort of underlying evil in [Being's] very positivity?" Levinas asks.

Yet, if the anonymity of being is evil, shouldn't the existent take up all the more cheerfully its existence? Shouldn't the existent get off the bed and shake off sleep with the willingness and the joy of someone who has just overcome, conquered even, the bonds of death? How do we explain then indolence, this lingering on the side of existence which is, at the same time, non-existence for it is not *my* existence? And so one might ask: if existence is perceived as evil, why does not the existent commit suicide? Levinas answer is that it is precisely this that the existent does by falling asleep. Going to bed and falling asleep is the closest one comes to committing suicide. My pajamas are my shroud.

Sleep is indeed comparable to death with the difference—as the analysis of insomnia shows—that death is not identified with the *il y a* but precisely with the escape from it. It is insomnia, the denial of that escape, which brings one face to face with anonymous being. Insomnia as a symptom of the death drive.

II. Radical alterity and radical evil

Levinas's insistence on the otherness of the Other is well known. It may be expedient at this point to rehearse the steps of this insistence in order to demonstrate that they inevitably lead to a radical alterity that opens up the possibility of radical evil. This assertion is particularly troubling to the extent that it may imply—perhaps against Levinas's intentions—an evil God.

There is no other portal to God available to us aside from the Other. "There can be no 'knowledge' of God separated from the relations with men," Levinas writes. "The Other is the very locus of metaphysical truth, and is indispensable for my relation with God."[26] This Other is not just another I, who is not like me, but an absolute other: "The absolutely other is the Other."[27] The otherness of the Other is not occasioned by a simple denial of the familiar, inasmuch as denial or erasure will indicate a gesture of negation, and every negation leads to a totality as far as "the negator and the negated ... form a system, that is, a totality."[28] The otherness of the Other, therefore, is not constituted by the eradication of any analogy between me and the Other (as in the case of via negativa); rather, it is safeguarded, Levinas says, by apostasy.[29]

The Other is not analogous to me but precisely the very opposite: disanalogous and asymmetrical. Asymmetry is the key characteristic of my relationship with the Other in its prohibitory function against the formation of any totality.[30] The face of the Other deprives the self of any autonomy; the self, as far as it is interpellated by the Other, appears to be in a heteronomical dependence on the Other. This disanalogous and asymmetrical relation with the Other is called metaphysics.[31] For Levinas, metaphysics, understood in contradistinction to ontology, is always a movement—not a circular movement, but a transcending toward the Other in a twofold fashion: transcendence (à l'Autrui) and trans-ascendence (à Dieu). Metaphysics, in this sense, radically opposes the scheme of an Odyssey where the self departs from its Ithaca in order to ultimately fulfill its nostos by returning to it.[32] Instead, metaphysics as transcendence toward the Other reflects the journey of Abraham as an exodus out of his fatherland to the unknown land where the imperative of the Other has called him. This "unknown land" is the Other, a "land" impossible to be ever possessed or appropriated, impossible to be inhabited or even seen, and yet one is ceaselessly and fervently vigilant in looking for it. What we are longing for, in other words, is this absolutely other, i.e., the Other. This desire for the Other, according to Levinas, is a metaphysical desire and it is absolute. "Desire is absolute if the desiring being is mortal and the Desired invisible."[33]

"Man as Other comes to us from the outside, a separate—or holy—face."[34] The "outside" of the Other, his exteriority (nudité) signals his resistance to be integrated into any system of sameness, that is, of a totality. The face of the Other is that which interpellates me and that which is beyond and exterior to any categories: beyond institutions, beyond comprehension, beyond any attempt of mediation. The Other's nudity constitutes a rapture of history; to the extent that the Other's face breaks with time in a movement beyond time, the event of the Other's appearance signals an eschatology. Levinas develops a notion of eschatology as the "relation with being beyond the totality or beyond history . . . a relationship with a surplus always exterior to the totality" and as "the breach of the totality, the possibility of a signification without context."[35]

Desire and eschatology: what brings these two notions together? This is how Kearney answers the same question: "for Levinas, both [namely, desire and eschatology] have the same ethical structure—namely, a relationship with the other."[36] Thus, the desire for the Other beyond totality, "a desire beyond desire," is an eschatological desire.[37]

Levinas suggests, rather boldly that a phenomenology of desire holds the key to this eschatological infinity. He thus confronts us with the paradox that the infinite is inscribed within our historical experience of totality precisely as a "trace" which betrays that which is "beyond" it. This trace, he argues, is evinced, first and foremost, by desire and responsibility–desire of the other and responsibility for the other. Eschatological desire is, in short, desire of the infinite.[38]

Kearney, in this reading, seems to disregard Levinas's emphatic and often repeated statement that "[w]hen man truly approaches the Other he is uprooted from history."[39] Perhaps one could say, without any injustice to Levinas's thought, that this uprooting, this rapture with history takes place within history, allowing thus for Kearney's reconciliatory interpretation.

Nevertheless, isn't it our face-to-face relationship with the Other analogous to our confrontation with anonymous Being? Could we not trace a certain parallel between, on the one hand, my desire for the Other under the imperative of such radical alterity and, on the other, the horror of facing the elemental? "Alongside of the night as anonymous rustling of the there is," Levinas writes under the subject "Phenomenology of Eros" in *Totality and Infinity*, "along the night as anonymous rustling of the there is extends the night of the erotic, behind the night of insomnia the night of the hidden, the clandestine, the mysterious, land of the virgin, uncovered by Eros."[40]

What could account for this equivocation between the erotic and the elemental? The steps undertaken in these last sections of *Totality and Infinity* are precise: first we are told that the beloved "presents a face that goes *beyond the face*"[41]; that the feminine constitutes an inversion and disfiguration of the face.[42] Not only it is stated here quite clearly that "Eros goes *beyond the face*,"[43] but also and more specifically that "voluptuosity aims *not* at the Other but at his voluptuosity; it is voluptuosity of voluptuosity, love of the love of the other."[44] Such a desire, however, that *trans*cends the face of the Other cannot but also *trans*figure him. Thus, "the beloved . . . has quit her status as a person,"[45] that is the mode of hypostatic existence and therefore, in accordance with the earlier distinction of hypostasis versus nature that is fully confirmed here, she is closer to the natural signified by the infancy of the child or even the animal "one plays with the Other as with a young animal."[46]

We are reminded at this point of the striking similarity between Levinasian erotics as we have described it here and Diotima's initiation to the mysteries of love from Plato's *Symposium*. Here too, as it was for Diotima, the beloved is little but a stepping-stone in the transcending and trans-ascending movement in the ladder of the beautiful.

Levinas's remarks that one is not in love with the Other but with the love of the Other or the good in the Other—with an abstraction in any case that moves beyond the specificity of the Other: his or her face—echoes faithfully Diotima's thinly disguised objection to Aristophanes' story and her affirmation that one is never in love with the half or with any part but rather with the good in the Other.[47] Yet, one would expect Levinas to disagree with the metaphysics of love of the *Symposium*. Strangely, though, not only he does not, as we begin to see here, but also the only criticism directed against the *Symposium* is precisely that story that one had hoped that Levinas would had embraced. So he writes: "the myth Aristophanes tells in Plato's *Symposium*, in which love reunites the two halves of one sole being, interprets the adventure as a return to self."[48] Levinas is aware of the possibility of this *egoism à deux* as he describes with patience and clarity the lovers' propensity to form an anti-social, egoistic society that succeeds in isolating them from the rest of the society.[49] He is not afraid, however, that such an accusation might be raised against him for he hastens to provide his solution in the "engendering of the child."[50]

And it is precisely the double function of the child that lies, in my opinion, at the root of this unlikely similarity between the *Symposium* and the last sections of the *Totality and Infinity*. In the *Symposium*, love is finally defined as the desire to be and to be forever. The secret of desire is the coveting of immortality. An immortality, however, which, since it cannot be achieved in the strict sense it soon becomes "trans-substantiated" in procreation and creation.[51] I cannot live forever but at least I can live, as it were, through my child. But who is exactly this I that enjoys this immortality by proxy? It is not, it cannot be the personal, hypostatic I but it is rather the impersonal I (and therefore no I at all) of my species, my genes, my nature. The undying, elemental nature that triumphs once again over hypostatic being. In fact, it is a double victory that nature has won against hypostatic being: my death and the loss of my specificity in and through my child, who, in turn, will have to continue feeding with its own being death's voracious mouth. What is achieve, if anything, by the fecundity of love is the continuation of impersonal being, that is, of the evilness of being.[52] This conclusion is, indeed, inevitable for a philosophical system which, like Levinasean ethics, refuses to ground transcendence to immanence, remaining thus vulnerable, as we shall see below, to the possibility of radical evil.

III. The ethics of desire

Although Lacan does not specifically use the term "eschatology" or "eschatological desire," he often employs a language that can be suggestive of the eschatological character of desire.[53] What gives us the right to speak of an "eschatological desire" are the references that, deliberately, I believe, Lacan makes in relation to a point of view that he calls "the Last Judgment." Take, for example, the following passages:

> [T]he question of the realization of desire is necessarily formulated from the point of view of a Last Judgment.[54]
>
> The experiment is adopting what I called the point of view of the Last Judgment. And I mean by that choosing as the standard of that reconsideration of ethics to

> which psychoanalysis leads us, the relationship between action and desire that inhabits it.[55]
>
> [A] reconsideration of ethics is possible, that a form of ethical judgment is possible, of a kind that gives this question the force of a Last Judgment: Have you acted in conformity with the desire that is in you?[56]

It becomes clear from the above passages that "the point of view of a Last Judgment" is undoubtedly what enables him to undertake a year-long (1959–60) reconsideration of the traditional ethics and which finally leads him to the foundation of a new ethics, that of psychoanalysis. It is also arguable that his programmatic statements in the beginning of the Seminar on the Ethics of Psychoanalysis remain out of sight throughout the length of his lectures until he reaches the example of Antigone. It seems that it is there, and by means of discussing the "Essence of Tragedy," that Lacan finds a new language, a new code, namely tragedy itself, that permits him to approach the ethical dimensions of psychoanalysis. And indeed, as soon as the discussion of Antigone is completed, he immediately addresses his real subject in the final section under the title "The Tragic Dimension of Psychoanalytic Experience." It is there that all the leitmotifs, elaborated in the previous sections, are orchestrated in a common synthesis. And it is there that the iconography of a last judgment is carefully depicted.

"Analysis is a judgment,"[57] Lacan says from the very beginning, but a judgment of what, on whom, and by whom? It is certainly a "judgment" that involves the subject's realization of his or her own desire and, through the desire as "the metonymy of our being," the realization of his or her own existence. There is something here that makes the "realization-of-one's-desire" so crucial to Lacan's understanding of psychoanalysis. He emphasizes the fact that "realizing one's own desire is necessarily always raised from the point of view of an absolute condition."[58] He even challenges his audience by saying, "Try to imagine what 'to have realized one's desire' might mean, if it is not to have realized it, so to speak, in the end."[59] And finally: "That is why the question of the realization of desire is necessarily formulated from the point of view of a Last Judgment."[60]

As such, the last judgment constitutes a final point, an absolute end of which one is living in anticipation of; what is characteristic of this final point is the fact that we are always introduced or led to it through a long period of expectation which, properly speaking, is an advent. During this long time one is living in anticipation, waiting; after all, the last judgment coincides with, or rather presupposes, a messianic coming. Waiting has always been the waiting for the other's arrival. At the moment of the coming and by the coming everything becomes open, visible, revealed: judgment finally takes place. The coming fulfills the advent; the long period of waiting is over and completed upon the coming. The Last Judgment thus signals the end—not a relative end, the end of this or that, but the end in its most absolute sense: the eschaton. This end is essentially the end of time, or better yet, the end of history.

So far this has been a reading faithful to Hegel, or at least to Kojève's lectures on Hegel's system. For such an absolute point as the end of history could not be realized unless through the obtainment of an absolute knowledge. The absolute knowledge is, according to Hegel, indeed possible and it is to be found embodied in the person of the wise man who, unlike the philosopher, does not desire wisdom any longer since he had realized and perceived himself as identical with the object of his desire: "knowing-with, finally, an absolute knowledge—that he himself is the object of his desire, the Hegelian Wise Man attains satisfaction (Befriedigung)."[61] There, where the desire of the man has reached its final pleroma—which means, it is fulfilled, completed and rested—there, there is nothing more to desire any longer, and therefore, history ends. This is what on Queneau's terminology we would call The Sunday of Life or, according to another tradition, the eighth day. As such, eschatology—the coming of God (messinanism) or the cunning of Reason (Hegel)—served as the "nodal point" around which History strives to find its meaning and thus to protect itself from the meaningless terror of the real.

Both Levinas and Lacan break at this point with the Hegelian dialectic. In Lacan's scheme neither the analysand nor even the analyst can declare any claim to absolute knowledge and thus to satisfaction.[62] The subject of desire cannot be, in any case, the same as the Object of desire, being granted that "the desire is the desire of the Other."[63] This Other, therefore, must remain within the boundaries of alterity. Thus, satisfaction (as jouissance) is ultimately unattainable in Lacan's system. Simply, satisfaction cannot be reached. For Levinas also, God qua Other (or the Other "as the manifestation of the height in which God is revealed"[64]) cannot be and should not be knowable, for any knowledge—and even more one that claims to be absolute—is a reduction of alterity to sameness, where the other ceases to be Other and, through comprehension and appropriation, becomes another I, an alter ego. In their effort to avoid a lapse into Hegelian grand closure, both Lacan and Levinas employ a different meaning of eschatology and eschatological desire. Lacan argues in terms of the "metonymic chain" that takes desire from signifier to signifier with no end, whereas Levinas describes the desire for the other as an insomniac night full of anticipation. Levinas is careful, however, to distinguish between two antithetical understandings of eschatology—namely, eschatological desire as the rapture of totality and eschatology as the triumph of totality (messianism). The latter, he says, is the view of history that constitutes "a finality, an end (fin) to the historical relation of difference between man and the absolutely Other, a reduction of the gap which safeguards the alterity of the transcendent, to a totality of sameness."[65]

Lacan's desire—as in the case of Tennessee Williams' *A Streetcar named Desire*—follows an endless route from one stop to the other but with no destination:

> I emphasize the following: the properly metonymic relation between one signifier and another that we call desire is not a new object or a previous object, but the change of object in itself.[66]

This allows us to cast a glimpse on the horrific image of the endless advent of desire. In every single moment of history, which is understood as a ceaseless succession of "desires" in the metonymic chain of signifiers (S1 . . . S2), a "difference" (the inexhaustible otherness of the Other, in Levinas's idiom) slips in between the subject and the Object of desire, rendering the attainment of desire's object impossible. Through this slippage, satisfaction is endlessly deferred; in other words, the-never-yet coming of satisfaction signals the over-coming of (this) desire for (another yet) desire.

In a similar line of thought, Levinas has described desire and the ethical responsibility that springs forth from it as "insomnia or wakefulness precisely because it is a perpetual duty of vigilance and effort which can never slumber."[67] Insomnia, wakefulness, and vigilance are all terms of waiting, stages of advent where one leaves always in anticipation of the Other's coming (regardless if this coming turns to be a Waiting for Godot, an infinitely deferred fulfillment, a permanently unattainable satisfaction).[68] What is interesting here is the paradoxical ability of eschatological desire to exceed the Other towards whom it is supposed to be always oriented. Levinas makes clear that the eschatological desire knows of no terminal. It floats toward the Other and, in a certain sense, even beyond him:

> Transcendence is time and goes unto the Other. But the Other is not a term: he does not stop the movement of Desire. The other that Desire desires is again Desire . . . Here the Desire which in the first pages of this work we contrasted with need, the Desire which is not a lack, the Desire that is the independence of the separated being and its transcendence, is accomplished—not in being satisfied and in thus acknowledging that it was a need, but in transcending itself, in engendering Desire.[69]

The question that presents itself at this point is to what degree the juxtaposition that we have attempted so far between the two movements of desire in Lacan and Levinas make it legitimate for us to speak of an analogous conception of the "desire of the Other," common to both these thinkers, as a desire that transcends towards an eschaton. There is, it seems to me, a distinction between two different species of otherness in Levinas and Lacan and, if so, two incompatible desires that are ordered, respectively. Levinas's Other is not completely overlapping with Lacan's Other. The latter's Other is the other that I see reflected on the glass of the mirror (as in the mirror phase), an inverted I, my double self. The first traces of Lacan's notion of the other are to be found in his doctoral thesis "On Paranoia and Its Relationship to Personality" (1932). There Lacan tries to establish the theory that his patient, whom he calls Aimée and who had knifed an actress, was perceiving her victim as an "extension" of her own self. By wounding the other (the actress), the self (Aimée) is punishing itself. One could also cite the similar example of St. Augustine's pleasure in living vicariously the pain of the tragic heroes on stage, as we have discussed it in the previous chapter. "Je est un autre," Rimbaud writes, and Lacan seems to have found a case that illustrates this claim. But then, what this will mean for my desire as "the desire of the Other"[70] insofar as this Other that I desire is an Other qua I? And what will be the implications for Levinasian desire of the Other qua God?

In the first reading of this key phrase, my desire as the desire of the Other qua I could be deciphered as "I desire my desire"—that is, I desire to desire regardless of the object of my desire, which ultimately is nothing but itself: that is why Lacanian desire is a metonymy and, at the same time, a paradox. "Je n' aurai pas la lune," says Caligula in Camus's play and explains, "si J'avais eu la lune, si l'amour suffisait, tout serait changé" (Act IV, Scene xiv). Maire Jaanus sketches the paradox of desire in the following way:

> It is Lacan's analysis of desire as a paradox that must itself provide proof of the fact that we do not really desire anything we desire; that we block getting what we want in order to maintain desire; that we lie, and abuse ethics, law, and illness to erect such blocks; that we will demand the most unobtainable things, such as God, immortality, or death, primarily in order to save our desire; and that we desire most steadfastly not only that which we can never have, but what does not even exist, which is therefore outside of existence, or nothingness itself.[71]

IV. The real God and the god of the Real

On the other hand, Levinas would have read the same phrase ("my desire is the desire for the Other") in a sense that this Other as the absolute other, far from being another I, moves all the way up in the axiological axis and trans-ascends to the heights. That is, he is invested with the status of God; then we should read: my desire is the desire of God. Given the intentional ambiguity of the double genitive (the desire of God). One could unfold this sentence in three alternative ways:

1. As a modality of infinity: the desire of God then indicates a desire for God; the loving relationship with God as it was experienced by the mystics of all the religious traditions in their visions and ecstasies. Levinas writes about it: "I can never have enough in my relation with God for he always exceeds my measure, remains forever incommensurate with my desire."[72] This kind of desire is prohibitory of any attempt to appropriation, possession, and comprehension. It opens up, in other words, a window to infinity, in front of which one humbly abdicates one's rights to power and its abusive repressions.
2. As a modality of totality: the desire of God signals a desire to see, to know, and to have God (the tripartite properties of *voloir* as *voir*, *savoir*, and *avoir*); a desire to possess God as the object of one's desire, or to be God, to be like God. This megalomaniac claim was aptly impersonated by Camus's *Caligula*, for instance as a masquerading "will to power," that is, a desire that wishes to immortalize us and make us omnipotent. It camouflages also a fear of death insofar as the desire to be God is the desire to be and to be forever (cf., the Symposium).
3. As a third option, finally, that refuses to succumb to any of these categories and that reads the phrase as the desire of God—that is, God's own desire, a desire proper to God.

This third option opens up a whole new problematic: is there any desire proper to God, and if so, what or whom does God desire? What is God's desire? To answer such questions amounts to posing the very question of God's identity, the question that asks who this God is and who he may be. Such a question, however, is *a priori* excluded by the system within which we are seeking it—that is, the Levinasian metaphysics. In Levinas's scheme there is no room for either knowledge of God or a relation with God. As we have already seen, any knowledge of God would seek to comprehend God in a manner where the par excellence Other will have to be reduced to the knowing self and well thus have to sacrifice the otherness of the Other/God to the demand of defining him and thus controlling him. A relation with God, on the other hand, seems to be less problematic, to the extent that Levinas welcomes it, but is such a relation really possible?

The relation with the Other is the locus where the Other presents himself to me; this presentation of the Other ignites my desire for the Other, which, as we have already seen, is an eschatological desire. The presentation of the Other's otherness to me takes place through and in the Other's face. The face-to-face relation with the Other is essentially the only possible relation with the Other. If so, how can one, any of us, have a face-to-face relation with God?

Have we ever encountered his face or could we ever see his face without losing our lives, since as a common topos in the Biblical literature specifies, "Thou canst not see my face, for there shall no man see me and live" (Ex. 33:20)? This is an age-old question, a dream saturated with agony, an anxiety that has haunted the prophet's vision: "Who may abide the day of his coming, and who shall stand when he appeareth?" (Malachi 3:2). A lesson that Levinas choose to ignore from his Judaic tradition is that the Most High might appear and call us through opaque symbols, hermetic traces, voices in the desert, soft breezes in one's sleep, dreams and omens—even through the flesh of a Nazarene man, the ultimate folly of the incarnation—but never in himself, that is, in his real appearance, by his real name. What Levinas asks for, therefore—the un-mediated, face-to-face encounter with the faceless—falls within the prescriptions of an alterity that becomes radical. There is a fatal element in the face of God.[73] The sight of God's face, if I were ever able to face it, would have been painful and fatal for me. In particular, Levinas's God is a bit too much other, too much asymmetrical, featureless, strange. Waiting for this God in the darkness of our insomniac nights is not just an advent full of anticipation, but could be an empty nightmare that can scare us to death. Thus God's face, if he has one, does not welcome a relation; it rather prohibits it. God as the absolute Other—tout autre—presents his radical alterity in a way that there is nothing to prevent us from perceiving it as monstrosity. As Kearney has aptly commented: "How can we ever recognize a God stripped of every specific horizon of memory and anticipation; how can we ever consent to a faith devoid of stories and covenants, promises, alliances, and good works? How can we fully trust in a God devoid of all names?"[74] And again: "Is there really no difference, in short, between a living God and a dead one, between Elijah and his 'phantom,' between messiahs and monsters?"[75]

Edith Wyschogrod is aware of all these problems, as she rehearses them in an attempt to defend Levinas in her monograph on *The Problem of Ethical Metaphysics*:

> A more troubling allegation is that if absolute transcendence is, of necessity, featureless and, at the same time, God is not to be identified with the human Other (a point explicitly stated), must Levinas not, faute de mieux, equate God with pure being, being without objects? This is an especially vexing problem because pure being, what Levinas calls the il y a, is described as the threatening nocturnal obverse of environments, of the elemental, earth, sea, sky, and the like, in which we are immersed and is the source of a mythical or pagan religiosity.[76]

Thus, the greatest risk the Levinasian God runs is to have his face fractured and collapsed under the pressure of all these prohibitions against language and signification, form, image, and incarnation. Such an unnameable, defaced, and iconoclastic God can be easily aligned with the malicious il y a, the nameless and evasive khora and finally, the blinding and painful das Ding. For the God of the Levinasian metaphysics cannot be mediated either by the Lacanian symbolic order or by the imaginary (both would amount for Levinas to an attempt of totalizing God's otherness); he remains, therefore, to be placed in the domain of the real, that is, the locus inhabited by das Ding. And indeed, that is the conclusion that Richardson draws in numerous places; in an article entitled "Psychoanalysis and the God-Question", he writes: "I am wondering, then, if the real God—the God before Whom David danced—may not be encountered in the Real,"[77] and in "Religion and Psychoanalysis" he clearly affirms that "an encounter with Him only takes place in the real."[78]

Richardson is not alone in these observations. In a paper on the relation of das Ding to the Other, Paul Moyaert follows a similar path of thought that takes him to this astounding conclusion: "God acquires demonic features." The demonization of God appears as a result of the fact that "God occupies the place of the Thing."[79] To adopt Moyaert's argument, it is "not the object itself but its suggestive symbolic value [that] regulates the desire's pursuit of satisfaction."[80] This fundamental principle indicates that no object with some (albeit the minimum) symbolic value can be the object of desire. "What is perceived as pleasurable or unpleasurable," Moyaert continues, "is always embedded in a subjective network of associative memory traces."[81] Fantasy, then, is also anchored in memory as the anticipation of the repetition, the imaginary replay of the "initial" pleasure recorded in memory. Both fantasy and memory cannot, of course, be understood outside the symbolic (hors-signifié). What, then, about God? What about the God that Levinas insists be located beyond symbolization, signification, and sublimation? This God would have been too much alienated (an alien God) to have any symbolic value and, therefore, to be the object of (our) desire. How can we love a God that we cannot articulate? And how can we desire a God that we cannot imagine? As such, God appears to be the object without qualities, about which, Moyaert adds that there is "something falling outside of the symbolic value of the object [which] receives excessive weight."[82] This "excessive weight" allots to God the place of the Thing, designating him as "a plenitude that is empty"[83]—a characterization that in the ears of the School masters would have sounded a lot like their understanding of prime matter.

God, as the inhabitant of the real, has no other (there is no other of the Other) and therefore, God has no desire whatsoever. God is unable to desire; he might generate our desire but he himself is empty of any desire and even more, of the desire to desire us. Gott-an-Sich, perhaps—that is a God devoid of any signification, imagination, language, remembrance, and expectation, the God/Good beyond or without the beauty of sublimation—is a violent God whose lethal presence would leave us breathless and lifeless.

Therefore, one could not but agree with Derrida's assessment that "the possibility of radial evil both destroys and institutes the religious."[84] The religion that radical evil institutes, and, in turn, is instituted by, is the religion inaugurated by the Kantian categorical imperative, which overthrows and replaces the divine law, and by its autonomous subject, who becomes independent from God.[85] It is Kant's natural religion that first proclaims the death of God even prior to Nietzsche, since the moral subject is supposed to act *as if* God did not exist. God, thus, becomes nothing more than a useless prosthesis and he can, in due time, be properly discarded. Derrida speaks of Kant's *Religion within the Limits of Reason Alone* as "a book on radical evil"[86] since it is radical evil that such morality presupposes without which "good would be for nothing" [*sans lequel on ne saurait bien faire*].[87] One, of course, detects here the Manichean duality that makes evil complimentary and necessary for the good.

Radical evil, as a possibility for religion, opens up in the positing of precisely such a Kantian God, that is a conceptual God, "universal" and, therefore, impersonal. For radical evil and conceptual God share the very same structure: that of radical alterity. A God who lacks a face and consists only in the dictation of the moral imperative is a God who prohibits or demands but cannot promise. It is only in promise, however, that there is memory, as the *anamnesis* of the promise, and hope, as anticipation for the fulfillment of the promised. Without memory and hope, without past and future, without the mediation of time and history, God becomes imprisoned in the now. Without memory and hope there can be in imagination—for there is no image of the immemorial or the unhoped for. Hence the iconoclastic prohibitions that we have encountered in Levinas's God of radical alterity as well as in Lacan's God of the real. Such iconoclasm is the result of radical otherness, which, in turn, opens up the possibility of radical evil. To re-state this conclusion, we could say that the absence of relation (with the Other) is the very root of the possibility of radical evil. An encounter with the divine without incarnation and without relation (to the extent that such an encounter might be possible at all) — that is, without some form of mediation, would not be any different from an encounter with evil.

The fundamental difference, therefore, between the real God and the God of the real is time: the God of the real is immediate in both senses of the word: without temporal mediation (history) and without the mediation of the face of the Other (intersubjectivity). In contradistinction to the God of the real, the real God welcomes mediation through time and embodiment as in the chief example of the incarnation.

It would seem, then, that Lacan was right in pairing Kant with Sade,[88] as was also Agamben right in tracing Levinas's genealogy back to the Kantian origin of radical transcendence.[89]

V. God in the cinema: on the Christology of *Dogville*

I believe that the foregone discussion could find an apt illustration in one of Lars von Trier's films: *Dogville*.[90] I would like, therefore, to engage in a reading of this film (in the same manner that we read Sophocles' *Oedipus Rex* in chapter 5), underlining those points that are pertinent to our analysis.

The very name of small town where the narration of the film is located and after which the film itself is named seems to be a witty anagram for "Godville"—the village of God. The fact, however, that the letters are put in *reverse* order might also indicate a God-less town. The only church in Dogville, a mission house, has never had a priest or sacraments performed there for, as Tom Edison Jr. says, "we can be spiritual without singing or reading from the Bible." In fact, the Mission House serves only as the location of his morally edifying lectures. If there is any religion in Dogville, that is only a religion "within the limits of reason alone." *Dogville*, nevertheless, could also allude to "Moses," the dog in Dogville, the only creature that survives the town's utter destruction; perhaps further evidence that religion here is limited only to the form of the moral imperative as in the Mosaic law. At the end of the film, Grace's father refers to the inhabitants of Dogville as "dogs" that "lick their own vomit" (an expression that alludes to 2 Peter 2:22 "Of them the proverbs are true: A dog returns to its vomit and a sow that is washed goes back to her wallowing in the mud.") The device employed by the director, such that in Dogville everything and everyone is always visible, since the walls of the houses and quite everything else is, in the eyes of the viewer, transparent, suggests a point of view that could be afforded only by God. The people of the village make a living by gathering and selling apples. The allusion here seems to be to the apple in the Garden of Eden—Dogville is an Edenic, post-lapsarian, town: living on temptation and living out temptation.

In that village, Grace appears as if she had fallen from heavens. The narrator says that Grace has come to Dogville as "a generous gift"—in fact, her name, Grace, means a gift (in theological language, grace is given as a gift by God). Upon her mysterious arrival, she is recognized only by Moses, whose bark indicated that he was "face to face with a force to be taken seriously." The first thing that Grace does upon her arrival is to steal Moses' bone, a suggestion, perhaps, of how the coming of grace signaled the passing of the Mosaic law. Grace herself says that she has "no family, only a father." A father, she adds, whose face she has seen. She is the only one who has seen her Father's face. Upon her arrival in Dogville, the town, we are told, undergoes "a tiny change of light." These "tiny changes of light" signal a change in perspective beyond perception, a moment of epiphany as the ones employed by Dostoyevsky in the narrative of *The Idiot*.[91] The music more closely associated with Grace's arrival—as well as a sequence of other important moments in Grace's life in Dogville—is always the same: the opening phrase from Pergolesi's *Stabat Mater* and Vivaldi's setting of Psalm 127 (*Nisi Dominus*, RV 608, in particular the four movement "cum dederit"). Both pieces betrayed the unmistakably religious, even liturgical overtones that underlie the narration of Grace's passion-story, namely the exploitation and abuse she suffered at the hands of the "good people" of Dogville.

The first person that Grace meets in Dogville is Tom Edison Jr. (his name of course is suggestive). He becomes Grace's supporter, advocate, and liaison with the village. Tom sees Grace's arrival as a unique opportunity for a pedagogical "illustration" for Dogville's need to learn how to receive and to accept. Tom Edison is something of a *moral* teacher for the town, a writer who never wrote anything except "*great and small?*" He is, in other words, a thinker who did not go further than the paradox and, therefore, unable, as we shall see, to understand the paradox that Grace is. His first words in the film are about what is "useful." We hear him speaking in a series of clichés (e.g., "it is the soil that gave us all life") that suggest a thought constrained by naturalism. But above all, he is a rationalist. He reasons about everything to the point that he is unable to act, like Hamlet in Shakespeare's play, even when a course of action becomes urgent and necessary. He is in love with Grace, if, indeed, he is capable of loving, yet he is unable to let Grace know of his feelings. In fact, the storyline of *Dogville* suggests that he is unable to love Grace (the particular person), for he values too much the universal, his own ideas. He watches silently as Grace is abused by each and every of his fellow citizens, unable to act, since he can only think. The only action he finally undertakes, as soon as he realizes that Grace poses a danger for his "moral mission" and "career" as a writer and thinker, is to betray Grace.

It is him who also comes up with the idea of Grace making her "gifts" available to the families of Dogville as a way of assimilating Grace in the everyday life of Dogville. So, Grace—according to Tom's plan—begins working for one hour for each of the eight families that live in Dogville. That amounts to a regular eight-hour workload per day. It is ironic, however, that Grace should perform chords that the people of Dogville, as they admit, don't "need." In return, Grace is to receive wages—thereby, inscribing the "grace" of Grace, her "gift," into an economy that turns gifts freely given into products or services. It is the law of the marketplace that prevails here: a *quid pro quo*. At the end, her presence in Dogville becomes "more costly," as Tom Edison puts it, speaking "from a business perspective" and, therefore, she has to work twice as much earning half a pay. Grace becomes the eyes to the blind man in the village, the leg to the crippled, and an extra pair of hands for everyone.

Soon, however, things change. Every single person abuses Grace, either physically (labor, rape), or psychologically (blackmailing, accusations, threats, etc.). Ultimately, Dogville will turn Grace, this heaven-sent gift, into a slave and a prostitute. As the film progresses we witness how the town of Dogville, a symbol of the entire humanity, commits one after another despicable acts of mistrust, greed, and hypocrisy. At the end, the village has *appropriated* Grace; they have turned a gift into a property that they now own and refuse to let go. She cannot escape Dogville. She becomes, quite literally, a slave and a prisoner of the "good people" of Dogville.

This seems to be the central question of the film: are the people of Dogville good after all? In the first town-meeting they boast saying that they are "good people," "knowing each other," "caring for human beings." In addition to Tom (the rationalist and moralist), we meet Vera with seven children, all of whom bear pagan names; she is an enlightened humanist (on the blackboard in her house one reads the words "Eros→Psyche") who, however, is capable of afflicting the cruelest punishment on Grace. We also meet her hard-working husband; the cheerful black cleaning lady with her crippled daughter; the good housekeeper; the naively religious adherent and so on. All simple, average, everyday people, as it would seem. And yet, if *Dogville* has a point this seems to be that, given the chance, poor people, people in need and misery can be—in fact, they *will* be—as mean and vicious as any other. That goodness is nowhere to be found—neither in the moralist, nor in the humanist, not even in the child. Each one of them (and in them we all recognize ourselves) becomes the very face of evil. Of course, they are evil in a very "humane" way for they reason—is rational evil less evil?—and provide excuses for themselves.

A number of subtle (and sometimes not-so-subtle) indications render this film as an allegory of the incarnation. To quote the prologue of John's Gospel "He was in the world, and the world was made by him, and the world knew him not. He came unto his own, and his own received him not" (John 1:10–11). Similarly, Grace leaves the "city" where she was with her Father, and she comes to Dogville only to be treated as a slave.

In the language of the Scripture: "being in very nature God, did not consider equality with God something to be grasped, but made himself nothing, taking the very nature of a servant, being made in human likeness. And being found in appearance as a man, he humbled himself and became obedient to death—even death on a cross!" (Phil. 2:6–8). *Contrary*, however, to the Christian account of the incarnation, Grace's coming to Dogville does not signal the salvation or the redemption that Christ brought to the world. Her coming to Dogville has a quite different outcome.

"The Final Illustration" (i.e., Grace's Judgment Day): so far Grace has had "the rare talent of looking ahead and only ahead." Her future-oriented outlook, one could indeed say, her eschatological understanding of the people and the town, enables her to see in the gooseberry bushes the fruit that they will bear "come summer." When Tom Edison Jr. introduced his fellow citizens to Grace, he appends each of his introductions with a negative comment. In the eyes of the moralist we are all obviously lacking. It is only Grace who can see them under a forgiving light. When, however, the people of Dogville, thinking that they could solve their problem with Grace by calling the gangsters and handing her to them, they bring upon themselves what they could have never expected—an unthinkable eventuality. Grace meets her Father again who now asks her to return "home" and be "like her Father," "sharing his power" and becoming "conspirator" (or "co-spirator"). Grace accuses her Father of being arrogant, for it is him after all that "passes judgment." The Father argues that it is rather Grace who, in her mercy, exemplifies arrogance for, in refusing to judge them ("they are human beings," as she observes with a compassion that sets her apart from the human nature), considers herself a standard that cannot be imitated. "Would she have not done the same like them, had she lived under the same circumstance?" she asks herself. Well, the matter of fact is that she has lived with them, like one of them, in the very same circumstances (if not worse) and she didn't act like one of them. The town undergoes another of these so telling tiny changes of light only now in favor of a present bereft of any future: "suddenly, you couldn't see the berry that will be there one day but only the thorn that was there *right now* . . ." At this very point, the logic of the incarnation takes an unpredictable turn—if humans can be judged (and indeed, they must), it is precisely because of the incarnation, because God became human and therefore humans have no excuse anymore for their behavior. It is the humanity of Grace ("for the sake of the human being that was Grace herself") that enables her to condemn them—and us—to utter annihilation. Every citizen of Dogville is violently killed and the village itself burnt to the ground.

Lars von Trier offers us an alternative interpretation to the mystery of the incarnation than what Christianity has traditionally come to understand in Christ's humanization. Instead of the possibility of a universal salvation, he suggests the possibility of a universal condemnation. The story, as we tried to indicate in this brief reading of the film, remains more or less the same; it is the conclusion that each reading draws from that story of God-becoming-man (or, in this case, woman) that becomes radically different. It is for this reason that we could say, without any exaggeration, that *Dogville*'s revision of the Gospel, by turning the "good news" of Christianity into a message of destruction, constitutes a masterful and perhaps the most serious threat articulated against Christianity to today—all the more so, since it does so by following as faithfully as possible the main theological tenets of the incarnation.

True enough, Grace as a Christ-figure is understood primary in moral than ontological terms and so *Dogville* reads more than a gospel written by Kant or Jefferson. Nevertheless, von Trier's cinematic interpretation of Christianity poses for us the crucial question of the difference between salvation and condemnation. Had the incarnation not taken place, one could have retorted that God could not pass judgment on humanity (does he, after all, know what does it mean to be a finite, mortal being?). Now, however, that God became man and dwelt among us, a human being in every respect like us *except sin*—how could we hope to escape judgment, that is, how could we not be found lacking when compared to him? How could we decide between salvation and condemnation when faced with the ambiguity of judgment?

Here the problematic of God's alterity, that is, the difference between the God of the real (i.e., the necessity of condemnation) and the real God (i.e., the possibility of salvation), is posed once again and with quite some force. *Dogville*'s distorted Christology rests upon a messianic figure (Grace) who is without mediation: unlike the Jesus of the Gospels, Grace is not born in Dogville; she has no mother to connect her with the human race and with the history of that race; she does not grow old—rather, she comes, or better yet, she *appears* (the risk of Docetism is quite obvious) in Dogville without any real connection to its citizens. She dwells in the town and among them, yet she remains other than them, an otherness unmediated and radical. Grace is a Christ without flesh and without history. *Dogville* tells the story of how the implication of such a messianic figure cannot be anything else than, necessarily, a necessary condemnation. Lars von Trier's *Dogville* reminds us that the subterranean connection between Kant's morality and sadism might be, indeed, inescapable.

There is nothing surprising about all this. In our discussion so far we have seen at length that without the change of mind—the *meta* of the *metanoia*, a change made possible through time, that is thought the reflection of the before from the point of view of an after—the good, contracted at the single moment, at a present without future or past, cannot but be perceived as evil. Now we begin seeing that such a contraction would have to render God evil as well. The Manichean hypothesis returns in and through modernity.

On the antipodes of Levinasian Manicheanism we discern the risk, equally dangerous, of what might be called a Heideggerian Pelagianism. The phenomenological analysis of temporality, oscillating between the hopeless and the hopeful, leaves the question of human agency unresolved. We do not naively suggest that evil, touched by the magic wand of time, will be transformed into the good, while all we have to do is wait patiently for the end of history. However, I think that we shall be wrong to understand the question of agency only or mostly in the limiting terms of activity, action, and activism. When it comes to the recognition of the good—which precisely as re-cognition entails always the reflection opened up by retrospection—agency operates on the level of passibility. That is, on our ability to undergo a passion in both senses: of affectation and of suffering. We can navigate the rough seas between Levinasian Manicheanism and Heideggerian Pelagianism by steering the middle course of an Augustinian grace of passibility—of the receptivity of the gift offered at the crisscrossing between temporal and intersubjective mediation.

The Third Garden: *Final Being*

9

An Undying Death

I. The self in hell

In writing about the last things and humanity's eschatological destiny we are immediately presented with the insurmountable difficulty of writing about what none of us could claim to have an experience and even, perhaps, about that which might fall beyond the scope of experience altogether, especially in the sense we understand experience phenomenologically. Nevertheless, as our comments in the prolegomena to this book have sufficiently explained, the eschatological constitution of consciousness has no difficulties in representing the not-yet in the now, neither collapsing the future into the present nor reducing the absent into a presence before me. The remarks that follow intimate the end of the journey we have been undertaking since the beginning of this book from our present position. Earlier, we were able to speak of the beginning—as we did in the first two chapters—and indeed of the absolute beginning, of creation, fully aware that neither do we have nor we could have an immediate experience of that beginning. Yet, this was made possible because the beginning reverberates through any act of consciousness. If this is so, how much more true should this be of the end, which, as the endpoint of consciousness' teleological orientation, sustains the possibility of experience as such.

On the other hand, we find ourselves confronted with a paradox. In writing about eschatology the difficulty consists not so much in not knowing enough, but rather in the excess of knowledge with which a plethora of diverse material makes available to us. All the difficulties in writing about eschatology originate in the fact that somehow we already know too much about it. Thus, the task at hand becomes to sort out through this confused body of knowledge and try to discover a way to organize and present it meaningfully.

For example, in approaching the idea of judgment and condemnation, the idea of hell, we are immediately presented with a wealth of vivid images that the poetic imagination, from Homer to Plato to Dante and beyond, has bequeathed to us through the literary canon of our tradition. Of such wealth of imagery we could say that Pluto (the name means wealth) was quite rightly the god of the underworld. These images can be summarized under two broad ideas common to them all: that of a hopeless existence of pain and suffering which is to last forever. In this chapter, we would like to discuss how these two ideas (hell's punitive character and hell's eternity) could be presented in a manner that is both philosophically acceptable and consistent with the phenomenological tenets of our foregoing analysis.

It would seem that the idea of suffering and the idea of eternity are intimately linked. So much at least is suggested by the familiar stories of afterlife punishment in Greek literature. The stories of Sisyphus, Tantalus, Ixion, and the Danaids (one could also add to this list Prometheus) offer unanimous witness to that connection. Even though their punishment in Hades takes different forms, the root of their torment remains the same: the futile effort, the attempt to satisfying a desire without success *over and over again*. It is time—but note well: a time identical in its repetitiveness—that constitutes their punishment. In Hades time does not move forward, in fact it does not move at all; it falls back upon itself in a semblance of movement that is nothing else than an everlasting present. Remove this element, and there is no punishment anymore.

Yet, being stuck in a time that is itself stuck in repeating itself is not a condition imposed on them externally. It is already found *in* the punished hero and comes from him; more precisely in the compulsion of repeating his desire. It would seem, then, that the gods of the underworld have left their victims to their own devices and done so for all eternity. Indeed, in the phenomenological description of sin in chapter 4, we described sin as "a tic" that "repeats itself almost compulsorily—although its compulsion is but an illusion facilitated precisely by its lack of continuity." We wrote then (and we see it confirmed here again) that "as a series of spasms do not make up a gesture, let alone a movement, so the repetition of sin over a given amount of time fails to make up the content it lacks." Removed from time, sin's static repetition emulates a bad eternity, such that we find underlining the punishments in Hades. Is it possible, then, to understand hell as a self-inflicted punishment? Kierkegaard thinks so, and he goes on to add more: not only hell is self-inflicted, hell is nothing more and nothing less than the self itself—a self that has infected itself with the sickness unto death.

Kierkegaard, writing under the pseudonym of Anti-Climacus, borrows the title of his work, *The Sickness Unto Death*, from the Gospel of St. John:

> Now a certain man was sick, Lazarus of Bethany, the town of Mary and her sister Martha... Therefore the sisters sent to Him, saying, "Lord, behold, he whom You love is sick." When Jesus heard that, He said, "This sickness is not unto death . . ."
>
> 11:1–4

Yet Lazarus dies. We know, of course, from John's narrative that Christ raises Lazarus in spite of the "stench" of his dead body that had been laid in the tomb "for four days"—thus, this sickness is not unto death. "But even if Christ did not resurrect Lazarus, it is not still true that his sickness, death itself, is not unto death?"[1] As Anti-Climacus remarks:

> When Christ approaches the grave and cries out with a loud voice, "Lazarus, come out" (11:43), is it not plain that "this" sickness is not unto death? But even if Christ had not said that, does not the mere fact that He who is "the resurrection and the life" (11:25) approaches the grave signify that this sickness is not unto death: the fact that Christ exists, does it not mean that *this* sickness is not unto death![2]

The structure of *The Sickness unto Death* is so systematic as to remind us of the fugues in J.S. Bach's *The Well-Tempered Clavier*: repetitions, counterpoints, chromatic developments, juxtapositions of themes, and so on. The definition of the self, which opens this work, is an example of such structure: "The self is a relation that relates itself to itself or is the relation's relating itself to itself in the relation."[3] The self is not the dialectic relationship between two categories, for example, between the infinite and the finite; it is not, in other words, a synthesis, nor is it the relation of such two categories to a third (as in the Hegelian triad of consciousness), but rather a second level of relation over and above the synthesis "of the infinite and the finite, of the temporal and the eternal, of freedom and necessity."[4] The self is a relation that relates itself to the relation between those pairs of categories, that is, self-consciousness. This is the first decisive step taken toward the understanding of despair. The second step concerns Kierkegaard's own position that this relation could not be self-founded, but rather it is found by another, beyond the human self (i.e., by God) and as such, self-consciousness is always a relation with a third, with him who constitutes it. "The human self is such a derived, established relation, a relation that relates itself to itself and in relating itself to itself relates itself to another."[5] Therefore, self-consciousness is the consciousness of the self's own heteronomic origin. This consciousness is realized in the recognition of a self that, by itself, is always lacking. These two initial positions will direct the understanding of despair as of two kinds: a) the despair of *not* willing to be oneself, and b) the despair of willing to be oneself. Both kinds, as it shall become evident in a moment, turn themselves against him who has grounded our selves as a relation—thus, the understanding of despair as sin treated in the second part of Anti-Climacus's work is already foreshadowed by this preliminary analysis. Finally, there is a preparatory stage of despair, the despair of failing to recognize oneself as a self, which functions as an introduction to the two main kinds of despairing. Let us then begin our reading of Kierkegaard's proto-phenomenological analysis from that last kind.

There is a particular kind of human being, whom Kierkegaard characterizes as "secular" or "natural" man and who lives according to the world.[6] This fact by itself is remarkable. And that because the self under no circumstances could be confined within the limits of the world, unless somehow it has become forgetful of itself and lost itself. If, on the other hand, today we fail to see this kind of person as remarkable, that is only because we are used to meeting almost exclusively these kinds of human beings and to such an extent that the paradox of this phenomenon, i.e., of worldly existence, has been attenuated to such a degree as to be taken as something natural. Of course, there is nothing natural in the natural man. On the contrary, as Kierkegaard's sagacious analysis presents it before our eyes, it is a disheartening spectacle:

> Surrounded by hordes of men. Absorbed in all sorts of secular matters, more and more shrewd about the ways of the world—such a person forgets himself, forgets his name divinely understood, does not dare to believe in himself, finds it too hazardous to be himself and far easier and safer to be like the other, to become a copy, a number, a mass man. (…) Just by losing himself this way, such a man has gained an increasing capacity for going along superbly in business and social life, indeed, for making a great success in the world. (…) As is natural, the world

> generally has no understanding of what is truly appalling. The despair that not only does not cause one any inconvenience in life but makes life cozy and comfortable is in no way, of course, regarded as despair.[7]

It is according to such a *sui generis* despair that the persons suffering from it are oblivious of their sickness and consider themselves healthy. In fact, the closer to death the healthier and more robust one appears. The natural man ignores that he is dead because he ignores that he can die. It is natural that the natural man should know of death only in its natural dimension, that is, only as biological death. *This* death despairs him, even though this death is not unto death, but rather his ultimate hope. "When the danger is so great that death becomes the hope, then despair is the hopelessness of not even being able to die."[8] The natural man does not know that he is a self and, therefore, he does not know that his self-ignorance has already brought about his demise. The kind of that man lives only according to the world, and according to the world he is just that: a desperate part of the world.

Anti-Climacus criticizes the mentality of the present *saeculum*, the secular mentality, for it identifies the infinite with the *indifferent* and confuses the one with the other: "for the secular mentality is nothing more or less than the attribution of infinite worth to the indifferent."[9] For the world, the indifferent—taken here also in its literal sense of that which ignores every difference, even the difference between God and humanity, between created and uncreated orders[10]—presents itself more and more in the garb of the infinite and, as if this was not enough, it usurps for itself the significance of the infinite: an infinite worth. Not even antiquity, despite all of its glorious achievements, seems to be exempt from that accusation:

> Every human existence that is not conscious of itself as spirit or conscious of itself before god as spirit, every human existence that does not rest transparently in God but vaguely rests in and merges in some abstract universality (state, nation, etc.) . . . every such existence, whatever it achieves, be it most amazing, whatever it explains, be it the whole of existence, however intensively it enjoys life esthetically—every such existence is nevertheless despair.[11]

The other two kinds of despair—the despair of not willing to be oneself and the despair of willing to be oneself—concern only those who are conscious of themselves as selves, that is, as being in relation with God. This relation, however, is not unproblematic, for a human being finds itself "thrown into the world" (*Geworfenheit*). The despair of the Christian comes from a double, and at times incompatible, experience, that of God and of the world. "If ye were of the world, the world would love his own: but because ye are not of the world, but *I have chosen you out of the world*, therefore the world hateth you" (John 15:19). Although being set apart from the world, this election ("I have chosen you out of the world") cannot but find the Christian *in* the world. The agony of the Christian despair stems from this double citizenship.

The despair of not willing to be oneself manifests itself as this imperceptible and manifold feeling of dissatisfaction. This sense of dissatisfaction does not make itself known immediately as a division within the self, but it first assumes the form of a variety of desires. Every time my desire is the desire to become someone else than whom I am. Anti-Climacus offers us a grotesque example: the man who got in his head to become Caesar, wants only this: "either Caesar or nothing."[12] Let us be careful here: the man who betrayed his dreams or who was betrayed by them, the man with the unrealized desires, does not despair over his failure, over the fact that he was not capable enough or lucky enough to realize his desires—this man has, above else, despaired over himself, for the self that he is. Desires, dreams, ambitions, all of these are not but the pretexts that bring to the surface a deeper despair—they lend to such despair a temporary form and name. If he was fooled in believing his own pretexts, this was only because in that way he could soon find a new dream, yet another desire in the chain of unrealized desires. Thus, we can say, together with Lacan, that what we have to do with here is not so much an unrealized desire but rather the desire of the unrealized. What one desires to become is precisely what one cannot and could not be, and what one dreams of is nothing else than what one could not obtain: a self rid of itself.

> This self, which, if it had become Caesar, would have been in seventh heaven (a state, incidentally, that in another sense is just as despairing), this self is now utterly intolerable to him. In a deeper sense, it is not his failure to become Caesar that is intolerable, but it is this self that did not become Caesar that is intolerable; or, to put it even more accurately, what is intolerable to him is that he cannot get rid of himself.[13]

The desire to become Caesar (or, closer to home, rich, beautiful, famous, and so on) is not but the concealed desire to become what I am not. Every such desire is suggestive of our flight from a self that leaves us dissatisfied and which assumes the form of a *phantastic* (and in the case where one wants to become Caesar, phantasmagoric) satisfaction of realizing some imaginary self.

Moreover, there are people who think that because they put on Caesar's clothes they have actually become Caesar. The real king, however, even when he is naked remains the king. I employ here only an example, which I believe the reader will be able to translate to cases more familiar to his or her everydayness. We have all met people who let themselves believe that their self originates from their "costume," their position, or their title. By fooling ourselves to believe that we are our roles we run a really inconceivable as much as terrifying danger for, to the extent that our self is identified with these "externalities" as Anti-Climacus writes,[14] it becomes all the more easy to strip ourselves from ourselves. The "king," therefore, that puts his royal clothes on, like an actor on stage, would lose much more than his costume when left naked. The nakedness of the naked king is painfully evident, first and foremost, to his own eyes.[15] It is for this reason that one is not quite sure what might be worse for such a man: to imagine himself as king or to become one?

Despairing over oneself is, for Kierkegaard, a sin: the denial of oneself is ultimately the denial my self's creator. Sin as the denial of God is always a refusal, that is, as much reaction as resistance over against the self's return to itself. Yet, since no matter how desperately one might try, one cannot cease from being oneself, damnation, that is, the experience of hell, amounts to the perpetual reminder that you are who you are: you are yourself. The self's very identity becomes its hell or, seen from a different perspective, its paradise. Odysseas Elytis's saying that hell and paradise are "made by the same material" and that they are different only in the arrangement of that material becomes particularly suggestive here.[16]

The despair of willing to be oneself, on the other hand, begins precisely from this point: from the moment that the self's identity has become an experience of hell. Descending to this self-made hell, or rather to the hell that we ourselves *are*, one cannot but feel desperate. Indeed, there is no hope as long as one remains within the limits and limitations of oneself; yet, hope (to which one does not even dare to hope) awaits us, provided that we are willing to take the crucial leap beyond ourselves. It is precisely this crucial step beyond oneself which a person inflicted with despair cannot undertake. He thus abandons himself to the hands of the sweet resignation that he could not possibly be anything other than who he is and this thought drives him to despair. In this case, however, his despair does not take the form of an anguish cry, but it is transformed into a defiant cry of triumph (or at least the tone that this particular kind of man adopts in order to cry out for help is that of triumph). If in the first kind of despair we recognized something unmistakably human, in this second kind we discern all the defiance of the demonic. Man's demonization consists precisely in his self-reliance, in refusing to recognize anyone else other than oneself. Indeed, the manic self-enclosure within oneself [*Indesluttethed*[17]], desperately clinging to oneself, and especially to one's sinfulness, is becoming only of the demonic.

Did it not this kind of man, this kind of despair and desperate rebellion, become the constellation formed by modernity's archetypical narratives? Did we not find in the imagery of the mythical Prometheus, of the occult Faust, of Leviathan, the paradigms of the modern man's rebellion which, with the words of Adam from *Paradise Lost*, raised his fist against his creator, saying:

> Did I request thee, Maker, from my Clay
> To mould me Man, did I solicit thee
> From darkness to promote me?[18]

The man in this kind of despair has a clear knowledge of his sickness. He knows well that his soul is infected by the sickness unto death and that, sick with an undying sickness, his torment will also have no end. Nevertheless, he resists with all his strength the proposed cure. He prefers to transform the whole of eternity into a protesting march. His sickness is the strength that enables him to stand up to his creator.

> Figuratively speaking, it is as if an error slipped into an author's writing and the error became conscious of itself as error—perhaps it actually was not a mistake but in a much higher sense an essential part of the whole production—and now this error wants to mutiny against the author, out of hatred toward him, forbidding him to correct it and in maniacal defiance saying to him: No, I refuse to be erased; I will stand as a witness against you, a witness that you are a second-rate author.[19]

Had he allow him to "correct" him, to heal him, to forgive him, would have amounted to recognizing the Creator as his creator, as his Lord. Yet, in the illusion of his autonomy, he cannot abdicate his self-mastery to anyone else other than himself. By himself, however, he cannot re-create himself (since he did not create himself in the first place). Thus, he decides to remain an error, going even further than that: loving himself as an error, "enjoying his symptom," becoming one with his sickness. What, after all, could stop a sickness that survives even death?

If Kierkegaard's discussion of the despairing self reads like an existential horror story this is only on one condition that is central to the outline of despair provided so far: that eternity had laid a claim upon the self. "Eternity is obliged to do this, because to have a self, to be a self, is the greatest concession, an infinite concession given to man, but it is also eternity's claim upon him."[20] This means that, for better or worse, one cannot escape oneself for all eternity. And indeed, we could add that in eternity such an escape would be impossible. For in this life, time and the body—that is, the body's temporalization of the self—provided the anguished self with many a distraction from itself. The descent to the bodily offered the greatest of consolations to such a self that now, left to itself, abandoned to itself, has no place to go, no place to escape to, because it has no place at all. Kierkegaard's sobering reading reveals a certain darkness in the light of the resurrection: in the end of all this, we come to realize that hell is a condemnation to life, a perpetual insomnia (to revisit a Levinasian theme). Humanity might have had condemned God to death, but God "retaliated" by condemning humanity to life. Humanity's dream to be like a god is finally granted, but that dream is proven to be a nightmare, for it is perhaps unbearable for a finite being, *as finite*, to be infinitely.

In chapter 7, we saw that the therapeutic moment for a self that suffers too much at the hands of its own narcissistic desires, its melancholy and anxiety, was exactly such a descent to the bodily—so much so that, in the absence of a physical trauma, the self has to invent one (or inflict one upon itself) in its desperate attempt to mitigate a pain that is precisely psychical and not physical, a pain that is bodiless. But from the perspective of eternity, if we are to follow Kierkegaard's silent omission of the eschatological body, such an expedient is impossible. Nothing, above all not God, could save such a self from itself, for if hell were only a place, we could be sure that God would have descended there in order to save the self from its torments. But when the self itself has become for itself its own hell, God is powerless for he cannot destroy what he has created and it is only such destruction, an absolute ontological annihilation that could save the despairing self from its despair. It is through this perspective that we come to grasp the reality of hell.

II. The second death

This existential vision of damnation should be supplemented with a theory already found in Origen and later developed and radicalized by Florensky. It is a theory that offers the hope of salvation at the very place where Anti-Climacus left us with the inescapable reality of self-condemnation. This unique theory resolves the much-debated dilemma between an all-forgiving God (universal salvation) and the self-condemnation of the self (universal damnation) by upholding both positions. As Florensky, in his characteristic style, writes:

> Thus, if you ask me, "Will there in fact be eternal torments?" I would answer "Yes." But if you were also to ask me, "Will there be universal restoration in bliss?" I would again say "Yes." The two are thesis and antithesis. (. . .) It is neither a simple "yes" nor a simple "no." It is both "yes" and "no." It is an antinomy.[21]

This antinomy finds, for Florensky, its theological correspondence and ground in the particularly Christian doctrine of the Holy Trinity. "The unity of God corresponds to retribution while the trinity of God corresponds to forgiveness."[22] Any preference for the one position over the other would introduce a serious distortion of the dogma of the Holy Trinity, either in favor of monotheism (universal damnation, God's just judgment, the hypothesis of *Dogville*), or in favor of Trinitarian communion (universal salvation, God's love, *apokatastasis*).

The antinomic unity of heaven and hell was first introduced by Origen in one of his early writings, the *Exhortation to Martyrdom*, and was further developed in his refutation of Celsus (*Contra Celsum*). In the *Exhortation to Martyrdom* we meet the idea of the self's division (διαίρεσις) and separation (χωρισμός) from itself, or at least from a part of itself.[23] Already at this nascent stage, this theory is invoking the scriptural passage of 1 Cor. 3:10–15 on which both Origen and Florensky will exercise their exegetical acumen. Later in the same work the idea of the coincidence of salvation *and* condemnation is articulated with the aid of a double metaphor: that of the word of God as a sword that cuts through the self's interiority (based on Heb. 4:12), and that of God as a purifying fire (attested in a number of scriptural passages). These two metaphors are then combined in order to support the theory that, at the end of times, the self in its duplicity will be "cut asunder" (διχοτόμησις[24]): all that is good in us will be preserved and thus saved; all that is evil will be consumed by fire. In this image, both heaven *and* hell, salvation *and* perdition is a common destiny awaiting every human being—they are both universal.

The passage from the *First Letter to the Corinthians* details the architecture of the human life: the whole scope of human endeavors is compared to the material we have chosen to construct our lives. Their value and truth will be tested only at the end, on the last day—if they shall endure the coming of the eschaton then they will abide, otherwise, when he appears "like a refiner's fire" (Malachi 3:2), they will be lost, but the human person in itself "will be saved" (1 Cor. 3:15):

> According to the grace of God which was given to me, as a wise master builder I have laid the foundation, and another builds on it. But let each one take heed how he builds on it. For no other foundation can anyone lay than that which is laid, which is Jesus Christ. Now if anyone builds on this foundation with gold, silver, precious stones, wood, hay, straw, each one's work will become clear; for the Day will declare it, because it will be revealed by fire; and the fire will test each one's work, of what sort it is. If anyone's work which he has built on it endures, he will

> receive a reward. If anyone's work is burned, he will suffer loss; but he himself will be saved, yet so as through fire.[25]

Origen offers the following interpretation on this Pauline passage:

> The divine word says that our God is "a consuming fire" and that "He draws rivers of fire before him;" nay, that He even appears as "a refiner's fire and a fuller's herb," to purify His own people. But when he is said to be a "consuming fire," we inquire what are the things which are appropriate to be consumed by God. And we assert that they are wickedness, and the works which result from it, and which, being figuratively called "wood, hay, straw," God consumes as fire. The wicked man, accordingly, is said to build up on the previously-laid foundation of reason, "wood, and hay, and straw." (. . .) And thus He appears as a "refiner's fire," to refine the rational nature, which has been filled with the lead of wickedness, and to free it from the other impure materials, which adulterate the natural gold or silver, so to speak, of the soul. And, in like manner, "rivers of fire" are said to be before God, who will thoroughly cleanse away the evil which intermingled throughout the whole soul.[26]

Origen's reading of the Pauline text was received by the Greek Patristic tradition under the authority of such thinkers as Gregory of Nyssa and Maximus the Confessor. St. Theodore the Studite, a ninth-century Byzantine monk, summarizes the long tradition of this interpretation in a passage from his *Letters* when writing:

> This is nothing else . . . than what the Apostle [sc., Paul] dogmatized, namely that the work of sin will be consumed by fire, while he [who committed the sin] will be saved unscathed by fire, yet not by participating to the good, but by having realized his deprivation of such participation. For what is from God, namely our nature, being originated from what really is [ἐκ τοῦ ὄντος], is indestructible by that fire. That, however, which is not from God, namely sin, having come to be by the sinner's choice, since it is not some being, but something accidental [παρυφιστᾶσα], it cannot be classified together with what is [τοῖς οὖσιν], and therefore it will be undone.[27]

At first sight, Origen's and Theodore's readings seems to lead to the well-known position of *apokatastasis*, that is, of the eschatological restoration of all, for which Origen was later condemned. However, this is not necessarily so. For one could raise two questions: firstly, "how much" of the self is lost as a result of this purification, provided that one's life may be entirely or mostly built by the material of wickedness? And secondly, how long does this trial by fire last—or rather for how long may the self that undergoes such a burning purification perceive this process as lasting? Florensky's treatment of eschatology offers a more detailed, though more gruesome as well, answer to these questions.

Based on Hebrews 4:12 ("For the word of God is living and powerful, and sharper than any two-edged sword, piercing even to the division of soul and spirit, and of joints and marrow, and is a discerner of the thoughts and intents of the heart"), Florensky agrees with Origen's basic idea of an eschatological catharsis, only for him condemnation takes the form of a separation that cuts the self asunder:

> Every impure thought, every idle word, every evil deed, everything whose source is not God, everything whose roots are not fed by the water of eternal life and is inwardly condemned because it does not conform with the Ideal which is in Christ and because it is incapable of receiving the Spirit—all this will be torn out of the formed empirical person, out of human selfhood.[28]

More specifically, what is thus separated is one's person (what he calls the "in itself") from one's character (the "for itself").[29] What is at stake in this "cutting asunder" is an ontological surgery of sorts, which finds its evidence in the eschatological parables of the Gospel (cf., Mt. 24:51, Lk. 12:46), and, at the end of which, one part of the self suffers a "second death,"[30] inasmuch as the separation of the in-itself from the for-itself repeats the separation of the soul from the body.

> After all, the human tradition has never ceased to keep this second death in mind by locating the end of our suffering there; in the same way it has never ceased to imagine a second form of suffering, a suffering beyond death that is indefinitely sustained by the impossibility of crossing the limit of the second death. And that is why the tradition of hell in different forms has always remained alive.[31]

This testimony comes from a rather unlikely source: these are the words of Jacques Lacan who, in his reading of *Antigone*, retrieves the notion of the second death as the paradox of a limitless limit. The psychoanalytic suggestion confirms Florensky's belief that the self that has been cut away from itself, in other words for the for-itself, does not perish but it endures in the pain of separation that persists forever. Thus he writes:

> If for the existent one, for "he himself," the eternity of torments consists in their *instant that is absolute according to content*, the instant when sin and God's gaze touch (and they cannot touch for more than an instant), then for the nonexistent [the "for himself"] this eternity is an unceasing continuation into bad infinity, an extension, absurd and a nonexistent-in-itself, of the dying of evil selfhood into the infinitude of purely inner, no-longer-restrained lust. (. . .) Left to itself, selfhood becomes it own slave, and, in the naked self-identity of the sinful I, it whirls senselessly like a vortex of dust, eternally impotent and never stopping in the darkness of nonbeing and torment. "Ye shall be as gods" has finally been achieved![32]

Here Florensky reaches the same conclusion as Anti-Climacus in the analysis of the despairing self, namely, the eternity of hell. The difference between the two is that for Florensky the same separation that confines the evil in me, or rather the evil me, to hell, it also salvages me *from* hell. In the passage above, Florensky stipulated that for one aspect of the self this ontological separation lasts only for an instant—the same instant which, for my condemned self is experienced as long as an eternity. There is nothing inadmissible in the suggestion that the same event or experience may be perceived, even by the same person, in the mode of two distinct temporalities. As we have already seen, the self's duplicity, that is, the distinction, however formal, between reflecting and reflected consciousness in Husserl; between the parasitic and fundamental self in Bergson; between Sartre's *être-en-soi* and *être-pour-soi*; between Levinas's existent and existence; between natural will and gnomic will in Maximus, and so on, renders Florensky's hypothesis not only plausible but also intelligible. The question, however, that immediately arises is: which part of the self in this list of distinctions is saved and which is condemned? By asking this question we return to the discussion that we have already undertaken regarding the eschatological destiny of the will in chapter 6. The same problematic applies here, for it would seem that for Florensky what is condemned is precisely that aspect of the self which distinguishes ourselves from others and which individuates our selfhood. "The entire content of consciousness will perish to the extent that it is not from faith, hope, and love,"[33] he writes in one instance, and he continues later in the text, "for the entire self-consciousness of wicked will is cast out, into the fiery darkness of the black and nonluminous fire of Gehenna."[34] What is, then, saved? The "person," as Florensky calls it, is the "image of God"—that which, in principle, cannot be destroyed, but also that undifferentiated characteristic which we all hold in common. Therefore, it would seem that for Florensky it is only the *human nature* that God saves, a faceless abstraction, but not particular individuals. Since we began this chapter with a discussion of Kierkegaard, it may be fitting to invoke at this point his warning that Christ did not come to save the human race, but rather to save us "out of the race."[35]

Finally, to the question whether the eschatological bliss of the saved will be marred, diminished, or rendered altogether untenable by the knowledge of the suffering undergone by the condemned (an argument which, for some, necessitates universal salvation), regardless of whether these are other people or parts of the separated self of each one of us, Florensky has the following surprising answer:

> Do we feel any sorrow about cut fingernails or even about amputated limbs? So, the righteous feel no sorrow about eternally burning selfhoods, which exist just as little for them as the unknown thoughts of other people exist for us.[36]

Here we need to pause for a moment, as Florensky's metaphor for the inconsequential suffering of my condemned self, inconsequential because essentially unknown and unknowable, is cast in terms of the familiar problem of *intersubjectivity*. The sorrows of those eternally burning selfhoods "exist just as little ... as the unknown thoughts of other people exist for us." We need to pause, therefore, in order to ask whether the question of intersubjectivity may not remind us of that which we have left out from this description of eschatology so far, namely the human body. And this because if "the unknown thoughts of other people" remain unknown, and so much so as to say that they do not exist for us, their bodies, however, not only exist and can be known by us, but, by means of the other's body, the other's otherness, the other as another consciousness, becomes revealed to me. Such was, after all, the solution proposed to the question of intersubjectivity in Husserl's fifth *Cartesian Meditation*. In narrating the travails of the suffering self, from Origen to Kierkegaard to Florensky, have we perhaps forgotten the body? The question of the forgotten body invites us to begin again our efforts to think the eschatological destiny of the human being: yet this time not merely as a self, but, more specifically, as an *embodied* self or, better yet, as an *animated body*. What, then, can we say of the eschatological state of the body? The question is particularly vexing as it presents us with the Gordian knot of two distinct and yet interrelated dualisms: on the one hand, that of body and soul and, on the other, that of time and eternity. Philosophy cannot wash her hands of the perplexities that these dualisms keep generating, for it was philosophy that created them in the first place—at least, since Plato's *Phaedo*.

III. The forgotten body: resurrection and the return to the beginning

Paradoxically, one of the texts that occupies a central place in the discussion of Christian eschatology is, at the same time, a text that most clearly demonstrates the appropriation of the Greek philosophical dualism by Christianity. I am referring to St. Gregory of Nyssa's influential dialogue *De anima et resurrection*.[37] One might indeed expect in a work entitled *On the Soul and Resurrection*—especially when styled after a Platonic dialogue[38]—to find Platonism's influence a little more pronounced than it is usually the case with the Greek Fathers of the fourth century. Nevertheless, to the extent that this is also a work *on the resurrection* and not only *on the soul*, it could have provided an excellent opportunity for a Christian to defend Christianity's kerygma over against the metaphysics of the Academy. Unfortunately, this opportunity was rather missed here.

One could begin justifying this assessment by noting that Christ's name is mentioned only a couple of times and even those are in passing. This silence is indicative, in my view, of a defective Christology, or rather the lack of any detectable Christological context. Christ's resurrection is indeed mentioned,[39] but no more than as the last example in a series of scriptural examples of resurrections.[40] It does not constitute any particular event in this discourse on the resurrection. I attribute to this work's absence of a robust Christology some, if not most, of the problems identified below.

Furthermore, the metaphysical structure within which Gregory's thought operates is, as expected, one divided between the sensible (τὸ αἰσθητόν) and the suprasensible or intelligible (τὸ νοητό). To these two orders Gregory gives his own particular coloring by further denominating the former as διαστημικό and the latter as ἀδιάστατον.[41] Yet, the mediation between these two orders is not the incarnate Word of God—never is the incarnation mentioned in this work—but the soul itself which participates in both orders by being on the one hand ἀδιάστατος and on the other κτιστή (i.e., created[42]). To make the soul a third term, a *tertium quid*, that is both created and yet without diastemic properties renders the Christological mediation between created and uncreated orders, as it will be defined later on by Chalcedon, superfluous.

This last point regarding the dual character of the soul requires a considerable explanation for in a number of other works, Gregory specifically restricts the character of τὸ ἀδιάστατον to the divine nature alone, that is, he makes it a characteristic of the uncreated (sc., divine) nature.[43] If the soul, however, is created, on pains of lapsing into Origen's error, then how is it also ἀδιάστατος?[44] A soul that is unlimited by space and time, as well as by any other diastemic characteristic, cannot but be only *one* soul. How are we to account for a plurality of souls, each of which retaining its individuality, in the absence of *diastema*? Perhaps the answer here is the body and the soul's connection to it—a connection which Gregory is willing to allow enduring even after death.[45] Yet, the question of identity and individuality, even though raised in most unambiguous terms by Gregory himself,[46] was never sufficiently answered in the course of this work.

In a sense, my objection is the reverse of that which Gregory voices to Macrina. If the risen body is not identical to and with itself then "one person will become a crowd of human beings."[47] Nevertheless, it seems more dangerous to me that, with every physical difference removed, a proposition that the resurrection as the return to the first creation[48] not only implies but also necessitates, then this crowd of humans, indeed the entire humanity, might collapse into one human being—or less than one.

Yes, Macrina indeed provides for one last difference: the ethical difference not of bodies, but of lives. Such a difference would distinguish between the various degrees of virtue and it would subsequently determine the manner of our participation in God's blessedness.[49] Yet, there is a further problem here: this difference in degrees of participation translates into a time difference (παράτασις), *for not all souls will participate to the eschatological perfection at the same time*. Hence the paradox that confronts Gregory. For, on the one hand, he asserts that there is *no time* at the eschata, yet on the other hand, the conformity of the wicked to the good will take *longer* as opposed to the almost *instantaneous* enjoyment of the eschatological bliss reserved for the virtuous. How do these temporal indications apply on a timeless eternity? This language of eschatological duration is more than a figure of speech employed in a slip of Gregory's pen. It is an argument of a central importance in his eschatological vision, as the following passage makes clear:

> When the complete whole of our race [παντός τοῦ τῆς φύσεως ἡμῶν πληρώματος] shall have been perfected through each man—some having at once in this life been cleansed from evil, others having afterwards in the necessary periods been healed by fire, others having in their life here been unconscious equally of good and of evil—to offer to every one of us participation in the blessings which are in Him, which, the Scripture tells us, "eye has not seen, nor ear heard, nor thought ever reached." But this is nothing else, as I at least understand it, but to be in God Himself; for the good which is above hearing and eye and heart must be that good which transcends everything. But the difference [διαφορά] between the virtuous and the vicious life will be illustrated in the future condition in this way; namely in the *quicker* [θᾶττον] or more *tardy* [σχολαιότερον] participation of each in that promised blessedness. For the *duration* [παράτασις] of one's cure will be analogous to the measure of the ingrained wickedness in each person. This cure consists in the cleansing of one's soul, and that cannot be achieved without an excruciating condition, as has been expounded in our previous discussion.[50]

The same objection can be raised against Gregory's celebrated notion of *epektasis*. Twice in the present work Gregory has Macrina ascertaining that the participation of the human nature to the good (i.e., to God) will expand "without ever ceasing to increase"[51] and it will be "without any limit."[52] The dynamism suggested by these expressions cannot be reconciled with the static *nunc stans* of a timeless eternity. These inconsistencies might have been resolved, had Gregory distinguished between two notions or kinds of the eternal: one reserved for God properly called *aeternitas* and denoting a lack of succession (ἀϊδιότης), while the other describing an endless time is more properly called *sempiternitas*. The same difference has been expressed by St. Maximus the Confessor as an eternity without neither beginning nor ending (ἀνάρχως) which belong to God and an eternity with a beginning but without end (ἀτελευτήτως) to which creation might aspire.[53]

Underneath this rather technical discussion hides the philosophical problem of *kinesis* (motion, movement) in St. Gregory's thought. The problem is connected with the metaphysical question of the relation between the one and the many, namely, with the effort to account for multiplicity without denying the ontological and chronological priority of the one. A clear expression of this question is found in Macrina's following words: "The question presents, on the face of it, many insuperable difficulties. How, for instance, could movement come from a nature that is at rest? How from the simple and undimensional [nature could come] that which shows dimension and compositeness?"[54] I am afraid that Gregory was not either willing or bold enough to vindicate *kinesis* by making it an intrinsic and inseparable characteristic of God's creation as such. The Eleatic suspicion against movement surfaces in such passages where Gregory speaks of multiplicity as the result of evil. The expression "by the entrance of evil [to creation] nature was divided into a multiplicity"[55] seems to justify the fears concerning Gregory's eschatology as the return (ἐπαναγωγή) to the undifferentiated *pleroma* of the first creation.[56] Such a return—if nothing more than a return to the beginning—would render Gregory's eschatology into archaeology. For when he says that: "the resurrection is nothing else than the *apokatastasis* of our nature to *the ancient condition* of our nature (εἰς τὸ ἀρχαῖον),"[57] he, in fact, establishes a circle where the final consummation is the Ulyssean return to the *archē*. And what was this *archē* to which we ought to hope to return?[58] Gregory explains, "at the *archē* the life of the human beings was μονοειδής."[59] Μονοειδής is rather an interesting term: borrowed from Plato's so-called argument of affinity between the soul and the forms (*Phaedo*, 78d5, and 83e2), it suggests a singularity of kind, it excludes multiplicity and differentiation—things which, as we have seen, Gregory attributes to evil—and it raises our suspicion that the eschatological return to the beginning might be taken to amount to a catastrophic, if it was not, in fact, *epistrophic*, undoing of creation's plurality, difference, and otherness. The *archē* is not only μονοειδής, it is also monotonous. In such circular schema there is little room for the dynamic movement of history, for a vision that vindicates movement and affirms time and the temporality of desire.[60] To this extent, therefore, and in spite of whatever modification he is willing to undertake, Gregory's view of history shows its weakness in resisting the allure of the origin—or, better yet, the allure of Origen.[61]

On the other hand, in Gregory's exegetical works the relation of the beginning to the end is regulated by the notion of *prolepsis* which is rather missing here. For example, in a passage from the commentary on the *Song of Songs*, Gregory understands creation's *archē* as a proleptic foreshadowing of the eschatological end—which is to say, that the beginning was only a preview of the end:

> When at the beginning [κατ᾽ἀρχὰς] the nature of creation was established by the divine power, for each of the beings' beginning was conjoined [τῇ ἀρχῇ συναπηρτίσθη] without any dimension [ἀδιαστάτως] to the end [τὸ πέρας] of all things that came from the non-being to being, having together with their beginning [τῇ ἀρχῇ συνανασχούσης] their perfection.[62]

The two terms employed in this passage—συναπηρτίσθη, "being cojoined," and συνανασχούσης, "having together"—spell out the connection between the beginning of creation and its end, making creation a *prospect* of the future perfection of beings so much so that their eschatological *apokatastasis* is not any more merely a return to the beginning as it is rather the completion of the unfolding of the end that began at that beginning.

These questions cannot be fully considered apart from the eschatological destiny of the human body. Embodiment and temporality are two inseparable aspects of one and the same phenomenon: of personhood. In a passage cited above, Gregory seems to suggest that a consequence of the fall was the division of the human life into two periods: a short time living in the body and an infinite period living *without the body*. Let us hear his own words:

> God divided the life of man into two parts, namely, this present life, and that *out of the body* [ἔξω τοῦ σώματος] hereafter; and He placed on the first a limit of the briefest possible time, while He prolonged the other into eternity [εἰς τὸ ἀΐδιον].[63]

Giving St. Gregory the benefit of the doubt (what did he mean by "body" here? How is it possible that he assumes the eternal life to be "out of the body" in light of everything else he says in defense of the body's resurrection), let us briefly take into examination the body's status in his eschatology. When, in the course of the dialogue,[64] Gregory raises the objection of the purpose that different parts of the body will serve once their function has ceased on account of the resurrection, Macrina's response is simply to reiterate that the resurrection will signal the reconstitution (*apokatastasis*) of our ancient nature,[65] which she avoids to describe in any detail, except for saying that "all of these [meaning such conditions as infancy and old age, and illnesses and "anything else of the bodily affliction"] were brought about by the entry of evil."[66] One wonders, then, how humans might have looked like in that first creation: without the need for food there would have been no digestive system, no mouth nor teeth; without the need for sexual procreation, there would have been no reproductive system; and so on (there is no need to be exhaustive here). What is left from the human body? What kind of body does he envision in the resurrection? And why, at this crucial moment, does he fails to look at the body of the risen Christ as the exemplar and promise of our resurrection? Indeed, Christ at his resurrection did not assume a perfect body that was blank of any traces of (his) history, but he appeared in that very same body, even down to the marks and wounds of his passion (John 20:27). It is in such moments that one feels more strongly this work's Christological deficiency.[67]

The reason for this deficiency is this work's pronounced Platonic atmosphere as witnessed by the many references to the dialogues of the great master of the Academy. Above all, however, what determines the problematic character of this work is the Platonic metaphysics of participation that frames Gregory's exposition from beginning to end. More specifically, the human soul is here an image (εἰκών) of the divine nature, both in its genesis as well as in its eschatological destiny. At the very beginning of Gregory's dialogue with Macrina the soul is conceived precisely in terms of bearing a *similitude*[68] to the archetype of divinity.[69] The same sentiment is reiterated in the dialogue's last words, only now the term εἰκών is extended to the whole human nature and not only the soul. The language of likeness (εἰκών) is of course corroborated by the foundational text of the creation of humankind in God's image (κατ' εἰκόνα ἡμετέραν, Gen. 1:26). It is not our task at the present moment to trace Gregory's extensive interpretation of this verse. Rather, what we are interested in is Gregory's adoption of Plato's so-called affinity argument (*Phaedo*, 78b4–84b8) at the expense of a more scriptural and Pauline understanding of the term εἰκών, reserved primarily for Christ as the "image of the invisible God" (εἰκὼν τοῦ Θεοῦ τοῦ ἀοράτου, Col. 1:15). One indeed wonders how much different a work the dialogue *On the Soul and the Resurrection* could have been, had Gregory chosen to steer a course closer to Paul than to Plato.

10

Sarx

Confiteor, quia omnes in hac carne resurgemus.[1]

I. Eschatology: with or without a body?

A century after Gregory's dialogue *On the Soul and the Resurrection*, the question of the eschatological body was still debated by the intellectuals of the now Christianized Roman Empire. A good example of this ongoing debate surrounds the life of Eutychius, Patriarch of Constantinople during the reign of Emperor Justinian. Eutychius presided over the Fifth Ecumenical Council, which condemned Origenism, and he later became involved in two theological controversies concerning the nature of Christ's body and, by extension, that of the human body in general. It is these latter controversies that are of interest to our discussion here. Following the Fifth Ecumenical Council, Justinian, who thought of himself as an expert theologian, issued in 564 a decree in support of the opinion that Christ's body was impassible (what is known as *aphthartodocetism*)—that is, immune to feeling any of the affectations of the human body, such as pain, weariness, hunger, etc. Justinian's motivations in the promulgation of such doctrine were mostly political: an incorruptible body, even prior to the resurrection, would have portrayed a Christ more sympathetic to the monophysite understanding of the incarnation, allowing thus the significant numbers of monophysites within the Empire to be reunited with Imperial orthodoxy. Yet in seeking to repair one rift, the Emperor-cum-theologian created another. Eutychius denounced the Emperor's degree and, as expected, he was dethroned from his see and was exiled.

Eutychius had recognized that positing an impassible body amounted to nothing short of denying the reality of the incarnation. If Christ's body was not subject to the same natural passions as ours, then salvation was only a semblance, an illusion. Then Christ was not really tired and thirsty when he met the Samaritan woman; he did not really cry when Lazarus died; he was not really afraid in the garden of Gethsemane; he did not feel any pain during his passion; and, finally, he did not really die on the cross. In the absence of a passable body, the incarnation, and everything that depends on it, becomes a badly acted farce. If Christ was unaffected by the human body, then the human body was unaffected by Christ.

Behind this obscure theological debate lay the old philosophical (Platonizing) contempt for the bodily. However, like Justinian who, though very energetic in condemning the spiritualization of Origenism, reinstated its principles on another level by his decree of aphthartodocetism, so Eutychius, who refused to accept the Justinian temptation of differentiating between Christ's body and ours, fell prey to the allure of Origenism, by denying the carnality of the human body after the resurrection. This second debate took place when he had returned from exile to the Imperial City and was reinstalled as its rightful bishop. In a dispute with a young delegate (*apocrisarius*) of the Roman Church in Constantinople, Eutychius defended the position that, after the common resurrection, the human body will be ethereal, "more subtle than air, and no more palpable"[2]—in short, not a body at all. The young delegate in the disputation with the Patriarch was none other than the future Pope Gregory the Great. In his great exegetical work *Moralia in Iob*, Gregory relates the details of this debate, taking as his opportunity the reading of Job 19:26 "et rursum circumdabor pelle mea et in carne mea videbo Deum" ("And I shall be again clothed with my skin, and in my flesh shall I see God").[3] "I hear of the resurrection," Gregory writes, "but it is the effect of the resurrection that I am searching out."

> For I believe that I shall rise again, but I wish that I might hear what kind of person; since it is a thing I ought to know, whether I shall rise again perhaps in some other subtle or ethereal body, or in that body wherein I shall die. But if I shall rise again in an ethereal body, it will no longer be myself, who rise again. For how can that be a true resurrection, if there may not be true flesh?[4]

For Gregory the reality of the resurrection is incumbent upon the carnality of the risen body. What is at stake here is the identity of two bodies, the body before the resurrection and the body after the resurrection, which, even though in some sense two, nevertheless they must be posited as identical otherwise, there "would be not a resurrection, but creation of a new human being."[5] Yet, at the same time as one seeks to affirm the identity of the risen body with that which was buried and decomposed, the two bodies cannot be the same, as if the resurrection itself made no difference, for then we could not speak of resurrection anymore but of mere resuscitation. As early as the very first attempts to articulate the Christian faith on the resurrection of the body, the difference between the mortal body and the risen body was emphatically underscored:

> The body is sown in corruption, it is raised in incorruption. It is sown in dishonor, it is raised in glory. It is sown in weakness, it is raised in power. It is sown a natural body [σῶμα ψυχικόν], it is raised a spiritual body [σῶμα πνευματικόν]. There is a natural body, and there is a spiritual body.
>
> 1 Cor. 15: 42–4[6]

Steering a middle path between, on the one hand, resurrection understood in terms of a new creation and, on the other hand, resurrection understood as resuscitation, requires the simultaneous upholding of both the identity and the difference of the human body. That is why it is rather unfortunate that the Pauline counterpoint cited above has misled some of the modern readers of this passage to assume that here Paul speaks not of two different modalities of the body, but of two different bodies altogether.[7]

The misunderstanding of such interpretation has been sufficiently corrected by James Ware who writes:

> The terms σῶμα ψυχικόν and σῶμα πνευματικόν function predicatively ("it is sown as a *soma psychikon*; it is raised as a *soma pneumatikon*") and thus describe two contrasting *modes of existence* of this same body, one prior to the resurrection, the other following the resurrection.[8]

If this were not the case, then the Pauline distinction earlier in the same text (1 Cor. 2:14–15) between two modes of existence for which the very same adjectives are employed, namely a "natural (psychic) man" [ψυχικὸς δὲ ἄνθρωπος] and a "spiritual man" [πνευματικός], would have made no sense if by them we were to understand two distinct and different individuals. There too, as well as in the passage above, the distinction is not according to quiddity or essence, but rather between two different modalities or forms of givenness of one and the same entity. "The spiritual man is therefore not distinguished from the psychic man, understood as a unity of soul and body only inasmuch as the first has something else that the second does not have."[9] Thus, what needs to be emphasized here is that the so-called spiritual man and the "natural" or "psychic" man "is the *same* man, identically drawn from the earth and breath (body and soul)," only the former "receives the Spirit of God."[10] Yet, he does so not in spite of his embodiment, but precisely *because of it*—for it is the flesh that, following Irenaeus, "is also the condition for the reception of grace."[11]

Indeed, this was the argument which Gregory advanced in his dispute with Eutychius when the latter invoked the same Pauline passage in order to support his belief of an ethereal, fleshless body after the resurrection by appealing to the Pauline warning that "flesh and blood cannot inherit the kingdom of God" (1 Cor. 15:50). Gregory's response, according to his own words, was as follows:

> In Holy Writ flesh is named in one way according to nature, and in another way according to sin or corruption. For there is flesh according to nature, as where it is written, "This is now bone of my bones, and flesh of my flesh" (Gen. 2:23). And, "The Word was made flesh, and dwelt among us" (John 1:14). But there is flesh according to sin, as where it is written, "My Spirit shall not always abide in those men, for that they are flesh" (Gen. 6:3). And as the Psalmist said: "For He remembered that they were but flesh, a wind that passes away, and comes not again" (Ps. 78:39). Whence too Paul said to the disciples: "But ye are not in the flesh, but in the spirit" (Rom. 8:9). For it was not that these persons were not in the flesh, to whom he was sending letters, but for that they had subdued the motions of carnal passions, henceforth, free through the efficacy of the Spirit, they "were not in the flesh." Therefore in respect to what Paul says, that "flesh and blood cannot inherit the kingdom of God," he would have flesh to be understood as applied to sin, not flesh as applied to nature.[12]

Gregory's (and Paul's) distinction between these two existential modalities of the bodily is of decisive importance to our discussion, as it anticipates the phenomenological distinction between body (*Körper*) and flesh (*Leib*). To discover this phenomenological distinction in the scriptural and patristic corpus is not an unwarranted attempt; Emmanuel Falque's phenomenological reading of the patristic tradition, from Irenaeus to Duns Scotus, is a painstaking study that confirms the conceptual alignment of the scriptural *sarx*, the patristic *caro*, and the phenomenological flesh (*Leib*), as distinguished from *soma*, *corpus*, and body (*Körper*) respectively.[13] It is this retrieval of a phenomenology of the body that will allow us to address and overcome the perplexities created by the double dualism between body and soul, time and eternity.

However, before we move on, let us return for one last time to Eutychius whose history of theological embattlements opened for us the question of the eschatological body. The last story we hear about him finds him in his deathbed; as various visitors come to pay their respects to the dying Patriarch, he used to take hold of the skin of his hand and solemnly pronounce: "I confess that *in this flesh* we shall rise again." His book on the resurrection, in which he had denounced the flesh of the risen body, was burnt and is now lost to us. His student and protégée, however, a certain Eustratius, presbyter of the Church of Constantinople, left us an apologetic work in defense of the life of the souls after death.[14] It was perhaps an attempt to restore his master's orthodoxy.

II. *Sarx*: the flesh of the soul

Under the influence of a certain misreading of Plato (more Pythagorean and Orphic than Platonic) one assumes that upon death the soul leaves behind the body and together with it time and history and enters into eternity. We cannot follow that hypothesis any further, for of such radically disembodied, incorporate, and eternal soul one could say, by definition, nothing more. Nevertheless, we could, following Methodius of Olympus, raise some objections against the hypothesis of an incorporeal soul. For a strictly incorporeal soul could not fulfill the purpose and function of a soul, inasmuch as it could not have any intercourse with the body, neither affecting the body, nor affected by it.

> If it [the soul] be incorporeal, must it not also be incapable of passion? For it follows, from its being incorporeal, that it is also impassible and imperturbable [ἀπαθὴν καὶ ἀπερίσπαστον]. If, then, it was not distracted by any irrational desire, neither was it changed by a pained or suffering body. For neither can that which is incorporeal sympathize with a body, nor a body with that is incorporeal, if, indeed, the soul should seem to be incorporeal, in accordance to what has been said.[15]

The argument for the soul's corporeality becomes intensified when transposed to what theologians called the intermediate state, that is, the time after the biological death and before the common resurrection. Such a state, with all the considerable difficulties it entails, becomes a hypothesis that tests and radicalizes one's understanding of the relationship between soul and body. The Christian praxis of the veneration of the saints and their relics, the offering of prayers for the repose of the dead, and, above all, the resurrection of the body had made the dualism of soul and body decisively untenable.

Stories of saints who appear to and intervene on behalf of the faithful abound in Christian hagiography. In what way, however, are the souls of the saints active within history? On the other hand, the central belief in the resurrection of the bodies, had raised the question of how could the soul resume the very same body to which it was united at birth after a long period during which the body had decomposed and vanished?

To the last question a number of authors maintained a natural bond between soul and body that persisted even after their separation upon death. For example, Gregory of Nyssa speaks of the soul's "natural relationship and love for the body with which it lived together." It is interesting to note that the language of cohabitation (συνοικῆσαν), avoids the image of the soul as living *in* the body, as if the body were merely a vehicle for the soul, rather both, soul and body, live together in the world. Gregory continues to say that "there exists in the soul a certain relationship and recognition [of the body as its own] as if there were some kind of traces [οἷον σημείων τινῶν] imprinted on the soul by nature, by virtue of which what is common remains unconfused, while what is particular is distinguished."[16] Without this bond with the body, embedded in the soul itself, a soul would have been undifferentiated and undistinguishable from every other soul—and this is an objection that confronts the Platonizing view of the soul, especially in light of the soul's reincarnation. In other words, under the assumption of the soul as essentially distinct from the body, the individuality of the soul must be sacrificed, which means, any argument for the immortality of the soul does not concern me, for I could not claim any soul as *mine* in a meaningful way. It is the body, *my body*, which safeguards individuality, this thisness (*haecceitas*) that eludes all description and discourse. The soul whose immortality ancient philosophy strove so hard to demonstrate, by distinguishing it so starkly from the body, is no one's and every one's soul. Ultimately, whether such a soul, without identity, is immortal or not is of no consequence.

On the other hand, Christian psychology recognized that, "even after the bodies are placed in the tombs, a certain bodily characteristic [γνωρίσματος δὲ τινος σωματικοῦ] abides in the soul."[17] There is, thus, an irreducibly bodily "element" in the soul on the basis of which, at the common resurrection, the soul will be able to reunite with its body by recognizing it, even if the body lies "in fragments."[18] This expression comes from the metaphor of a fragmented piece of pottery which one is still able to recognize even when broken. So the soul, Gregory argues, will be able to recognize what belongs to it. To speak of the dead body's "fragments" (λείψανα) here is significant, for the same term denotes a saint's relics. Perhaps a justification of the veneration of relics is here intended. For the soul "remains even after dispersion the guardian of its own."[19]

Similarly, John Scottus Eriugena, following Gregory of Nyssa, maintains that at no point is the soul disembodied—even after death:

> Though by corporeal sense they [body and soul] seem to be separated, when we consider this in a deeper sense, we see that it is necessary that they always, at once, and inseparably, subsist together. For the human body, whether it is alive or dead, is the body of a man. In the same way, the human soul, whether it is governing this body in unity, or ceases to rule it, as it appears to the senses, when the body is dissolved into parts, nevertheless it does not cease to be the soul of a man. And

> thus, it is given to understand by a higher understanding of things that the soul rules the body no less when it is in the dispersed elements, than when it was united to it by the complex of its members.[20]

For Eriugena, the relation between body and soul "is never able to perish."[21] This implies that he understands this relation as ontological—note his insistence that the body is "the body *of* a man" and the soul is "the soul *of* a man"—inscribed within an ontology of relation, where relation is not merely an accident of being, but rather what constitutes being as such *and* as the kind of being it is.

The most daring statement yet for what Gregory called above a "bodily characteristic" of the soul comes from Methodius of Olympus who, writing against Origen, did not hesitate to call the soul a "body," albeit "spiritual." It would be indeed difficult to translate with precision what Methodius meant in calling the souls *σώματα νοερά*, or when he wrote that the soul is *ὁμοειδὴς τῇ σαρκί* ("of the same kind as the flesh"). Methodius's "sarcic soul" sounds as paradoxical as Paul's "psychic body" and demands the same kind of attention with which we approached the Pauline distinction between a psychic body and a pneumatic body in 1 Cor. 15: 42–4. Even though, as we have already seen, by this distinction Paul does not answer the question of *what*; that is, he does not imply two essentially different bodies or, even less, two distinct individuals, but rather, he answers the question of *how* ("how will the body rise at the eschaton?") and, thus, his answer describes two different modalities according to which the same body is manifested; nevertheless, the distinction between a psychic body and a pneumatic body relies on a different kind of a distinction that remains hidden, namely the distinction between the now of time and the then of eternity. Methodius bold language of a soul that is both sarcic and somatic offers an invaluable service to our discussion by complicating further the clear-cut Pauline distinction.

We remarked earlier that the shorthand designations *sōma* and *sarx* correspond roughly to the psychic body and the pneumatic body, by anticipating the phenomenological differentiation between *Körper* and *Leib*. However, it would be a mistake to think, under the influence of 1 Cor. 15, that upon death or at the moment of the resurrection we exchange the one for the other. Methodius's language blurs the Pauline distinction: by calling the soul *σῶμα νοερό* (a "spiritual" body) while, at the same time, affirming that the soul is sarcic ("of the same kind as the flesh"), he conflates the *now* of the psychic body with the *then* of the spiritual body (since the "spiritual body" is the soul, *psyche*, and, so, the psychic body as well). If Paul's discourse on the resurrection refused to incorporate the Hellenizing dualism of body/soul,[22] Methodius's sarcic soul refuses to follow the Hellenizing dualism of time/eternity. As the psychic body and the pneumatic body of 1 Cor. 15 are not two different bodies but two different modalities of givenness according to which the one and the same body manifests itself, similarly the now of time and the then of the eschaton are not two different "times," but rather two different modalities of givenness according to which the moment of the present manifests itself. Or, to put it differently, *sarx* is the nodal point of embodiment and temporalization, the chiasmus of soul and body, time and eternity. After all, it is the lack of the flesh that prevents one for undergoing the process of *meta-noia*, that is, it is thanks to the flesh that the self is open to the effects of the good's temporalization as John Damascene remarks: "for it is thanks to the body's weakness that a human being comes to have repentance."[23] If an evil angel is not susceptible to *metanoia* that is only due to its incorporeality.

III. The flesh in the place of *das Ding*

"Omne corpus fugiendum esse": this saying, attributed to Porphyry (Augustine, *De Civitate Dei*, XXII. 26) suggests that, in order to attain happiness, the soul must flee the body. I would like to make a bold assertion here and say that such "flight from the flesh" is indeed *impossible*, insofar as the body itself *is* this "flight from the flesh," from a flesh, moreover, to which one is always bound to return. For both psychoanalysis and phenomenology the soul, in its most profound and fundamental understanding, which is also the less understandable, is *somatic* or, better yet, *sarkic*, that is, flesh. I have in mind here Freud's observation about the "somatic influences" of the id[24] and Husserl's distinction between, on the one hand, a *body* (*Körper*), organized by language, and, on the other, the in-articulable flesh (*Leib*).[25] Flesh is that which remains unsaid and unsayable, more intimate to me than my body, yet irreducible to any form of signification or reflection. The flesh as *l'impossible á dire*,[26] as the intimacy of exteriority, as *extimacy*, occupies the place of *das Ding*.[27]

The following is an effort to inscribe the flesh, as the place (and in the place) of *das Ding*, within a conceptual genealogy that connects Heidegger to Kant, Marion to Heidegger, and both to Lacan who, in his effort "to return to Freud," moved beyond Freud, insofar as the "return to Freud" took the form of a Husserlian "return to the things themselves" or at least to the "thing itself," *das Ding an sich*. The Kantian influence on Freud is noted by Lacan when he writes: "That is something which emerges in the philosophy of someone who, better than anyone else, glimpsed the function of *das Ding*, although he only approached it by the path of the philosophy of science, namely, Kant."[28] Kant's categories, namely quality, quantity, relation, and modality, provided Marion with the index according to which he organized the phenomenon of saturation. The event overflows quantity, the idol quality, the flesh overcomes relation, and the icon resists modality.[29] So Marion writes:

> If we follow the guiding thread of the Kantian categories, we locate, according to quantity, invisible phenomena of the type of the event (collective or individual); according to quality, phenomena the look cannot bear (the idol and the painting); according to relation, absolute phenomena, because defying any analogy, like flesh (*Leib*); finally, according to modality, phenomena that cannot be looked at, that escape all relation with thought in general, but which are imposed on it, like the icon of the other person *par excellence*.[30]

Here the flesh takes its place as one of the four paradigmatic phenomena of saturation. What is the relation of the flesh as the unrelated to the other three saturated phenomena, to the event, the idol, and the icon? Doesn't it the flesh constitutes the possibility of the event, like my birth and my death? Doesn't it the flesh give rise to the body, mine and the other's, which bedazzles me like an idol? Finally, isn't it the flesh the harbinger of an alterity inscribed deeper than my ego, quite literally "in my flesh"?

IV. Metaphors of death and resurrection: the pouring jug, the broken jar, and the eucharistic chalice

Heidegger's 1950 lecture under the title *Das Ding* provides us with a tentative answer to these questions. Heidegger's guiding question in this lecture is the seemingly simple question: "what is a thing?" In search of an answer, Heidegger takes up the example of a jug. To know what a jug is, it would seem enough to know that it was made—made by a potter, made by earth, made for the purpose of containing, made in the form of a jug.[31] Thus, Heidegger rehearses Aristotle's four causes: efficient (the potter), material (the earth), final (the task of containing), and formal (the form of a jug). Yet in a characteristic Heideggerian twist, Heidegger re-writes Aristotelian causality by calling attention to "the emptiness, the void" of the jug. "The empty space, this nothing of the jug, is what the jug is as the holding vessel."[32] In creating the jug, the potter does not create a thing, or rather it creates a thing only to the extent which, in a manner of a creation *ex nihilo*, "he shapes the void. For it, in it, and out of it."[33] Thus, "the vessel's thingness does not lie at all in the material of which it consists, but in the void that holds."[34] We could similarly say that the body's corporeality does not lie at all in the material of which it consists (the body as object), but in the void that holds (the body as flesh).

The void of the jug, however, is a void *saturated with givenness*. For "to pour from the jug," Heidegger continues, "is to give." "The nature of the holding void is gathered in the giving."[35] "The jug's jug-character consists in the poured gift of the pouring out." "In the gift of water, in the gift of wine, sky and earth dwell. But the gift of the outpouring is what makes the jug a jug." So, in conclusion, Heidegger arrives at a new definition "the pouring jug occurs as the giving gift."[36]

Marion's indebtedness to Heidegger here is beyond question. It should be added that Heidegger's quest to re-think the thingness of the thing is framed by a discussion on nearness and distance along the very same lines we find in Marion's study from 1977 *The Idol and Distance*. It is in this book, as well as in its companion *God Without Being*—both quite noticeably under Heidegger's influence—that the terminology of idol and icon, one of the two pairs of saturated phenomena, was first developed. Furthermore, the fourfold of saturation—the event, the flesh, the idol, and the icon—is Marion's own attempt to re-write Heidegger's now famous *Geviert*—earth, sky, mortals, and gods—as it was first developed in this lecture, "Das Ding."

In their respective endeavors, both thinkers, Heidegger and Marion, attempt to re-think *reality*. Yet one could not hope to make much progress in this attempt without confronting, or rather without being confronted, by the thing. After all, reality is above all the realm of the *res*. In Greek reality, *πραγματικότητα*, is the domain of the thing, of the *πράγμα*. Heidegger, in the same lecture we have been discussing, offers a thoughtful lineage of the various transformations and transmutations which such words as *res* and *causa* underwent through history. For us here, however, it is enough to note that the secret of the reality-principle is held by the thing, by *das Ding*,[37] insofar as it comes to signify an exteriority by far more alien to the subject than the external world which is, at the same time, more intimate than the subject is to itself: hence, Lacan's term *extimacy*.[38]

The entirety of the psychoanalytic project has often been summarized in Freud's phrase: *Wo Es war, soll Ich werden*—which, in another language and in another narrative, was known as "*ubi tu, ibi et ille*," that is, "where [your] mother is, there shall you be."[39] Yet, how could one occupy the mOther's place? That is, what does it mean to speak of the flesh as the place of the *das Ding*? What are we thinking in speaking of a place in relation to *das Ding*?[40] Can we assign *it* a place, unless, of course, this phrase: "the place of *das Ding*" is nothing but an infelicitous metaphor, a manner of speaking. Yet if for one fleeting moment, we were to take ourselves seriously, wouldn't we be surprised by the implication of what we are saying without explicitly saying it—wouldn't we come to realize that what is left unsaid, yet inescapably implied, in speaking of "the place of *das Ding*" is that *it*, as the thing par excellence, insofar as of all things it alone is only a thing and nothing else, a thing and nothing more, a thing without qualities or attributes, as "a plentitude that is empty,"[41] as an excess of nothingness, then *it*, if it is to take place, must be something bodily, if not the flesh itself?

I would like now to turn, by way of an illustration, to Pirandello's story *La Giara*.[42] The story is not about Heidegger's jug but about an oil jar. It is the story of a miscarried attempt to repair—one could even say "heal"—a crack that inexplicably has split the jar into two halves. A tinker is called to repair the broken jar and we are presented with two possible solutions. One is a "miraculous" resin cement that the tinker himself has invented; the other is the more traditional method of opening holes along the crack and holding together the two halves of the jar by applying rivets. The owner of the jar, Don Lollò, prefers the latter method.

In the process of mending the broken jar, the tinker rivets himself inside the jar—"imprisoned, imprisoned there, in the jar he himself had repaired."[43] Don Lollò, fearing that freeing the tinker could only mean breaking the jar once more, decides to keep him there, in spite of his lawyer's warning that such an action would amount to "illegal confinement,"[44] unless, of course, the tinker is willing to pay him for the value of the jar. But this is the problem: for the tinker the jar is already broken and, thus, it has no value, at least not the value that its owner would wish. So the tinker refuses to leave his confinement and remains within the jar.

There are some points of interest for our discussion of this story. The confinement of the tinker within the jar came as a result of the owner's insistence. As one of the farmhands observes: "the man on top gives orders . . . and the man on bottom is damned!"[45] The comment is about Don Lollò, the man on top, and the employed tinker, Dima Licasi, who, as the man on bottom—and soon to find himself on the bottom of the jar quite literally—follows orders. Yet, we wonder whether the man on top and the man on bottom might not be the same man. To say this is not only to pay homage to Heraclitus's saying "the way up and the way down is one and the same" (Fr. 60), but to think of the artificiality—Plato's noble lie—of such stratifications whether in the *Republic* or in the dynamic operation of the psyche. There is no secret about the secret alliance between the super ego and the id on the ego's expense. Seeing it under this light, the tinker's confinement inside the jar is self-inflicted.

More importantly, the central idea of this story seems to be that somehow a short, ugly tinker who, as we are told, looks "like the old stump of a Saracen olive tree,"[46] comes to occupy the place of the jar's void. The complication, the story's complex, if you wish, is that a man looking like an olive-tree takes the place where olive oil should have been. A metonymic substitution: the olive tree for the olive oil. "Oil oozes out," as Don

Lollò observes, anticipating perhaps the filling of the jar with something that cannot simply been poured out.

Finally, something about the jar itself. By becoming confined within it, the jar becomes the tinker's outer shell, a thicker skin and an extension of his body, *a second body* (Körper). We said that he is imprisoned there by Don Lollò's insistence to follow his orders, that is, by the "top man" in him and, in spite of his initial demands to be liberated, the imprisoned tinker, decides to "take up residence in the jar."[47] He becomes identified with this second body—the body as idol in Marion's terms. If Dima the tinker is to come out—if the olive-tree looking like man is to "ooze out" like oil—then, he has to be pressed, crashed like an olive in order to flow like oil.

The Brothers Taviani included a cinematic rendition of this story as part of a film structured by five short stories by Pirandello under the common title *Kaos*. The title is interesting as it allows us to make a connection between the film and Lacan's reading of the Heideggerian jug as a paradigm of creation *ex nihilo*: from chaos to cosmos. "And that is why the potter, just like you to whom I am speaking, creates the vase with his hand around this emptiness creates it, just like the mythical creator, *ex nihilo*, starting with a hole."[48] In more than one way, Lacan's three registers are organized in the image of the jug, with the symbolic and imaginary orders structured around the nothingness of the real, as means of both coping with it and protecting from it. *Das Ding*, then, occupies the place of an excluded fourth, at once exterior and interior to the subject, and the origin of an original trauma that allows us to consider the manner in which *the other* is constitutive of my subjectivity.

> The reason is that *das Ding* is at the center only in the sense that it is excluded. That is to say, in reality *das Ding* has to be posited as exterior, as the prehistoric Other that it is impossible to forget—the Other whose primacy of position Freud affirms in the form of something *entfremdet*, something strange to me, although it is at the heart of me . . .[49]

Finally, to Heidegger's jug and to Pidandello's jar, let us add a third example, that of the Eucharistic chalice. It is by means of a trauma, like the pressing of the grapes and of olives, that the nothingness of the void inside the chalice is filled with the wine that is offered both to God, like Heidegger's libations, and to all who participate in the Eucharist.[50] The pouring out of the Eucharistic chalice presupposes a previous *kenosis*, that of Christ's self-emptying. It is by virtue of that self-emptiness that the word became flesh, and subsequently this flesh became a body wounded and crucified, a broken body, which will be continuously broken, like the oil-jar around the same Dionysian celebration in order to release the flesh from its self-confinement inside its bodily jar.

At the culmination of this celebration, the celebrant invokes over the void of the chalice the Holy Spirit, re-enacting thus the original and originary moment of creation *ex nihilo*, when the Spirit of God hovered over the chaotic abyss. Creation begins again, endlessly, as a gift that keeps giving itself in excess of the flesh.

Epilogue: Three Scandals (In the Style of Dostoyevsky)

I. The scandal of evil

God's encounter with the devil in a Judean desert is well known (Lk. 4:1–13; Mt. 4:1–11). What the devil offered Christ through these three temptations was nothing else than a project of humanitarianism: bread, power, and authority. Feed the poor: that is have compassion for all the unjust evils that humanity suffers; bedazzle the crowds with the ingenuity of your miracles: that is, satisfy humanity's need for progress, let them believe in their infinite capacities. And, finally, give them someone whom they can follow and obey.

However, this was not the first encounter between God and the spirit of humanitarianism. The same temptations were posed to God a long time ago in the Garden of Eden. There, too, the devil—a sincere philanthropist—approached God and asked him: Why don't you make this world perfect? Why don't you eradicate all evil? Is it that you can't? Then, you are not God. Is it that you can but you don't want to? Then, you are not good. There, too, God turned a deaf ear to the devil's temptations and the devil moved on to the next step. You see, the devil's concern over the existence of evil was quite genuine—after all, who knows better about evil than him who suffers it daily? So he went to man and said to him: Make the world a better place. Where God has failed you can become, in his place, a god. Correct his mistakes. You are his best creature after all, the most gifted, the most capable. Correct the old Man's mistakes in his name and on his behalf. And man, moved so much out of love for God and his creation, humbly accepted the spirit's proposal and undertook the noble cause of making this world a better place. From that moment on, the devil became humanity's guardian spirit—the spirit of humanism.

II. The scandal of the good

Many centuries later, the philanthropic spirit had to step in once more and save the world or, as it was rather the case, save the world from committing a most terrible evil. It was a day full of turmoil on the foot of a hill in Jerusalem when the spirit of humanism ventured his most daring deed. Humanity was about to commit the most atrocious act in its history: the killing of an innocent man. And only humanity's spirit understands how every man is really innocent. But this was a particular case—somehow it had gone into this Man's head that it was his mission to go ahead and die, offering himself, as if it were, as a sacrifice not to humanity's progress or betterment (no, for *that*, humanity's spirit would have not only understood but even helped with all its power), but as a sacrifice to God. Humanity's spirit saw immediately as in a moment of flash, the tremendous consequences: on the one hand, the death of this Man was an unnecessary evil—man's goal, so the spirit thought, is the pursuit of happiness: the joys of conjugal life, growing old with one's beloved, watching one's children grow, living life in the fullest. The death of this Man was a cacophony in this dream. On the other hand, his unnecessary death would become an evil for those who would carry out the heinous crime. They, too, must be prevented from committing such an act of evil. But the spirit's greatest concern lay on another point: should this Man go on and die under the assumption that his death was willed by God, then a terrible injustice—indeed, a blasphemy—would have been brought against God, for God desires no man's death, especially not the death of an innocent man, and isn't every man in some sense innocent?

The spirit remembered with terror another instance, a terrible accident that took place not far from here. It was another murder, that of a son at the hands of his own father, and this too was to take place under the false assumption that God had so willed it. The spirit recalled how it had averted that murder at the last minute, providing an animal to be substituted for the innocent boy.

So it seemed that the same mistake was to be repeated once more. The spirit of humanity approached this Man as he was about to be stretched on his cross. He whispered most clearly to him how God loved him, for God is only love and therefore he could have never allowed one of his children—yes, one could even say, his Son—to die. His suffering was enough. What was about to happen was an unfathomable evil and if one had any compassion for humanity would do well to stop it.

Yet, somehow, nothing is more evil than to imagine Christ growing old.

III. The scandal of grace

> Now there is in Jerusalem by the Sheep Gate a pool, which is called in Hebrew, Bethesda, having five porches. In these lay a great multitude of sick people, blind, lame, paralyzed, waiting for the moving of the water. For an angel went down at a certain time into the pool and stirred up the water; then whoever stepped in first, after the stirring of the water, was made well of whatever disease he had. Now a certain man was there who had an infirmity thirty-eight years. When Jesus saw him lying there, and knew that he already had been in that condition a long time, He said to him, "Do you want to be made well?" The sick man answered Him, "Sir, I have no man to put me into the pool when the water is stirred up; but while I am coming, another steps down before me." Jesus said to him, "Rise, take up your bed and walk." And immediately the man was made well, took up his bed, and walked.
>
> John 5:2–9

What about the rest of the "great multitude of sick, blind, lame and paralyzed"? Why healing one when having the power to heal all? Would it not have been better if he hadn't healed even that one, if he is to overlook the many who now would have to endure not only their suffering but the fact that they were overlooked, neglected, ignored? Indeed, how would they have felt being left to their ailments, but more importantly being left behind, in a waiting without end and without hope?

Ah, the incomprehensible logic of exception, the cruelty of God's mercy! Which sensible humanist would not protest to him, protest of him, of that "merciful God" who heals one man while he leaves behind a multitude of sick people suffering? What sense does that one healing make?

No. It is not a God that we need. We are better off with a health care program. A system that would make sure that no one is exempted or exceptional, that no one is left behind. Either cure them all or else leave them all in their suffering: that's, after all, *justice*. Far from us, in any case, the injustice of the one, the exceptional, the exemplary! Our sensibility will not tolerate inequality. Even if it is to become an equality among sick people, blind and lame and paralyzed, at least it will be an equality, a democracy—that's the right word—of people happy and proud of their sickness!

Ah, the unbearable scandal of exception. One! Who can tolerate that? One saved or one savior, it is all the same. Our humanism cannot tolerate either. We rather believe that all should be saved and by all, or none by none. But if there is to be salvation it cannot be the salvation of the individual. Let it rather be the salvation of the nation, of the race, of all and by all.

* * *

In a world and for a self that are not perfect—for they could not be perfect as long as they are still in movement, moving toward the end, as long as they are still in-between the beginning and the end—shouldn't the good be perceived and experienced necessarily as evil? Furthermore, in a word and for a self content with their imperfection, an imperfection accepted as normal and normative, how else could the good, itself a perfection, be received and experienced unless as a trauma, a wound, an evil?

Isn't it this what Paul calls precisely "the scandal of the cross" (1 Cor. 1:23)? Wasn't it the cross the good perceived as the good's very opposite, as radical evil? And isn't it, therefore, every good a cross or, like the cross, a scandal?

Notes

Chapter 1

1 Plato's definition of time as "the moving image of eternity" (κινητόν τινα αἰῶνος … εἰκόνα, *Timaeus*, 37d5–7) is famous. In Aristotle's *Physics* Δ (IV.12) time is defined as "a measure for movement" (μέτρον κινήσεως, 221a) and "number of movement" (ἀριθμὸς κινήσεως, 221b 11). Plotinus is in agreement with the Platonic definition of time as a moving "image of eternity" (*Enneads*, III.7.11), adding that the temporal movement was first engendered by the Soul. This Plotinian gloss was influential on Augustine's conception of time, even though their understanding of the soul is different.

2 These two levels are examined under the name of two of their representatives in Ricoeur's masterful treatment in *Time and Narrative*. Thus Aristotle stands for the cosmological approach, while Augustine represents the psychological formulation of the problem of time. As Ricoeur writes: "The aporia of temporality, to which the narrative operation replies in a variety of ways, lies precisely in the difficulty in holding on to both ends of this chain, the time of the soul and that of the world. This is why we must go to the very end of the impasse and admit that a psychological theory and cosmological theory mutually occlude each other to the very extent they imply each other" (*Time and Narrative*, vol. 3, p. 14). Our efforts here are directed to "holding on to both ends of the chain" and see the one in the light of the other.

3 See, for example, Clare Carlisle's *Kierkegaard's Philosophy of Becoming: Movements and Positions* (New York: SUNY Press, 2005) and David J. Kangas's *Kierkegaard's Instant: On Beginnings* (Indianapolis: Indiana University Press, 2007), as well as Shannon M. Nason's work in *Motion, Change, and Activity in the Thought of Søren Kierkegaard* (unpublished thesis, Purdue University, 2008).

4 Polycarp Sherwood, *The Earlier Ambigua of Saint Maximus the Confessor and his Refutation of Origenism* (Rome: Pontificium Institutum S. Anselmi, 1955). See in particular pp. 92–116. Maximus's contribution on the conception of *kinesis* has been summarized in the reversal of the Neoplatonic (and Origenistic) triad *stasis-kinesis-genesis* into *genesis-kinesis-stasis*. The reversal takes place in Maximus's *Ambiguum* 7 of which more below.

5 S. Kierkegaard, *Concluding Unscientific Postscript to the Philosophical Fragments*. Edited and translated by Howard V. Hong and Edna H. Hong (Princeton: Princeton University Press, 1992), 253.

6 Parmenides, DK. 5.

7 Parmenides, DK. 8 (my translation).

8 Parmenides, DK 3.

9 "And motion is truth and not semblance, as Anaxagoras proves in spite of Parmenides by the indubitable succession of ideas in our thinking." F. Nietzsche, *Philosophy in the Tragic Age of the Greeks*, translated by Marianne Cowan (Washington, DC: Regnery Publishing, 1962), 92.

10 "*Nous* is the cause of motion [τῆς δε κινήσεως αἴτιον εἶναι τὸν νοῦν]. For Anaxagoras says this: 'When *Nous* began to move [things], there was separation off from the mutlitude that was being moved, and whatever *Nous* moved, all this was dissociated' and as things were being moved and dissociated, the revolution [περιχώρησις] made them dissociate much more'" (Fr. B13; Patricia Curd, *Anaxagoras of Clazomenae* [Toronto: University of Toronto Press, 2010], 25). "*Nous* in Anaxagoras's theory is a source of motion and change, and the principle of intelligibility in Anaxagoras's universe" (Ibid., 144).

11 Plato, *Phaedo*. 97b–98b. Translation by G.M.A. Grube in *Plato: Complete Works*, ed. By John M. Cooper (Indianapolis and Cambridge: Jackett, 1997), 84–5.

12 In spite of Anaxagoras's support of the ἄπειρον regarding the ingredients of the original mix (their number?), nevertheless, *nous* knows them—or perhaps it would be better to say that they are known in the directedness of *nous*. "For he [i.e., Anaxagoras] makes it clear that he thought that they [the things being separated off] were limited in form when he says that *Nous* knows them all. And indeed if they really were unlimited, they would be altogether unknowable; for knowledge defines and limits the thing that is known" (Fr. B 7); Curd, *Anaxagoras*, 21.

13 What is still missing is what one finds in Kierkegaard later on: not merely the movement of time that carries with it the world (and humanity), but the self-movement of a will in becoming oneself. See Clare Carlisle's *Kierkegaard's Philosophy of Becoming: Movements and Positions* (New York: SUNY Press, 2005), especially p. 62. More on this distinction below.

14 F. Nietzsche, *Philosophy in the Tragic Age of the Greeks*, 107.

15 Representative of this mistake is Gregory Vlastos's "Reasons and Causes in the *Phaedo*" in *The Philosophical Review*, vol. 78, no. 3 (1969), 291–325. So also Patricia Curd commending on Socrates' reference to Anaxagoras in the *Phaedo*: "Socrates is convinced that all good explanations are teleological, but he does not explain why this must be so. (. . .) Moreover, insofar as it lacks a teleological element, Socrates' hypothesis about Forms as causes or explanations in the *Phaedo* is introduced as a 'second best' theory" (*Anaxagoras*, 136, note 20).

16 That, in spite of all his talk of movement, Heraclitus's view falls short of affirming change, I find Clare Carlisle's following remark particularly insightful: "Because the aesthete has no passion, his inner world resembles Heraclitus's *cosmos*: without any fixed points or solid ground, everything is true and so nothing is true; everything is in motion and no*thing* is in motion." In *Kierkegaard's Philosophy of Becoming: Movements and Positions* (New York: SUNY Press, 2005), 56.

17 Dionysius (the pseudo-Aeropagite), *On Divine Names*, V.8.824C: "Παραδείγματα δὲ φαμεν εἶναι τοὺς ἐν θεῷ τῶν ὄντων οὐσιοποιοὺς καὶ ἑνιαίως προϋφεστῶτας λόγους, οὓς ἡ θεολογία προορισμοὺς καλεῖ καὶ θεῖα καὶ ἀγαθὰ θελήματα . . ." (*De Divinis Nominibus* in *Corpus Dionysiacum I*, Beate Regina Suchla (ed.), Berlin: De Gruyter, 1990, 188).

18 Prior to Dionysius, St. Gregory of Nyssa had borrowed Plato's analogy and terminology from the second sailing in *Phaedo*. Thus, whereas Socrates thought that "ought to be careful that I did not lose the eye of my soul; as people may injure their bodily eye by observing and gazing on the sun during an eclipse, unless they take the precaution of only looking at the image reflected in the water, or in some similar medium" (*Phaedo*, 99e), Gregory advises us that "man does not possess a nature that would be able to look steadily into the God Logos as into the sun-disk; rather he looks at the sun in himself as in a mirror. The rays of that true and divine virtue shine forth

from his purified life by the riddance of passions, which emanates from them, and make the invisible visible to us and the inaccessible comprehensible by displaying the sun in the mirror that we are" (In Cant. 3, GNO VI, 90).

19 Maximus the Confessor, *Ambiguum* VII (1080B), *On Difficulties in the Church Fathers*. Translation Nicholas Constas (Dumbarton Oaks Medieval Library 28, Cambridge: Harvard University Press, 2014), 97.

20 One such instance is found, for example, in his autobiographical poem *De Vita Sua*, 1153–4 in which Gregory refers to "those who acknowledge nothing divine but movement alone,/by which this universe comes into being and is set in motion" (translation by Caroline White in *Gregory of Nazianzus: Autobiographical Poems*, Cambridge: Cambridge University Press, 1996, 95–7).

21 *Oration* 43, PG 36, 521: "πάντα ἐν πᾶσι κεῖσθαι." The anonymous scholiast traces the phrase as well as the cosmological doctrine expressed by it back to Anaxagoras (PG 36, 911).

22 "And Anaxagoras by saying that mind [nous] is the cause of order and harmony for beings he was not far from our [i.e., Christian] beliefs. What else did the Prophet David indicate than this when he said 'by the word of the Lord the heavens were made and all the host of them by the breadth of his mouth' [Ps. 33:6]?" PG 36, 826.

23 That importance is testified by Kierkegaard himself; see *The Concept of Anxiety*, edited and translated by Reidar Thomte (Princeton: Princeton University Press, 1980), 242 (hereafter abbreviated as CA).

24 The connection between *Philosophical Fragments* and *The Concept of Anxiety* is made on several occasions by these two texts. On a note on p. 90 of CA Kierkegaard alludes to *Repetition* by writing "[h]ere the category that I maintain should be kept in mind, namely, repetition, by which eternity is entered forwards." Thus, a trilogy of sorts is formed by *Philosophical Fragments*, *The Concept of Anxiety* and *Repetition*.

25 The monotheletic controversy was a Christological debate that emerged sometime early in the seventh century concerning the question whether Christ had one or two wills, divine and human, corresponding to his two natures. The position acknowledging only one will (from which the debate draws its name) was condemned by the sixth ecumenical council (Constantinople, 681). Maximus was one of the chief opponents of the monotheletic error.

26 For a discussion of Maximus's contributions to the philosophy of the will, see my essay "The Dialectic of Communion and Otherness in St. Maximus' Understanding of the Will" in *Knowing the Purpose of Creation Through the Resurrection*, edited by Bishop Maxim Vasiljević (Sebastian Press, 2013), 159–81.

27 S. Kierkegaard, *Concluding Unscientific Postscript*, volume 1, Howard V. Hong and Edna H. Hong (trans.) (Princeton: Princeton University Press, 1992), 96.

28 For Origen's adaptation of the theory of the soul's pre-existence, see his *De Principiis*, III, 3 (English translation of Koetschau's text by G.W. Butterworth [Gloucester, Peter Smith, 1973], 228). However, Origen adopts the pre-existence of the souls with some modifications due to his Christian sensibilities, thus, the souls are not unbegotten but created by God (I, 3, p. 30), yet they predate the human body and they can go through a succession of embodiments not all of which are necessarily in human form (I, 8, 72–4); it was on account of such pre-existence that the souls fell away from God, a falling described as a process of losing their old fervent love, thus growing cold [*psychesthai*], from which they derived their names [*psyche*] (II, 8, 123–4). These views of Origen, and perhaps even Origen himself, were condemned by the 5th Ecumenical Council convoked by Justinian in Constantinople in 553

(see *Conciliorum Oecumenicorum Generaliumque Decreta*, ed. by G. Alberigo [Brepols, 2006], 161 and 183).

29 Kierkegaard, *Philosophical Fragments*, edited and translated by Howard V. Hong and Edna H. Hong (Princeton, Princeton University Press, 1985), 10 [hereafter abbreviated as PF].

30 Kierkegaard, PF, 181 (*JP* II 2088 [*Pap.* II A 448] May 29, 1839).

31 Kierkegaard, PF, 187 (*Pap.*V B 404:8, *n.d.* 1844).

32 It is tempting to suggest Plotinus as one of the Alexandrians. Plotinus did, indeed, spend some time in Alexandria where, together with Origen, he studied with Ammonius Saccas, but his association with that city was interrupted on account of many journeys and his final sojourn in Rome where he taught. On the other hand, the two Christian Alexandrians, Origen and Clement, are both named in a list found in Kierkegaard's notes from Henrik Nicolai Clausen's "Dogmatic Lectures" (see Paul Martens's "Origen: Kierkegaard's Equivocal Appropriation of Origen of Alexandria" in *Kierkegaard and the Patristic and Medieval Traditions*, volume 4 of *Kierkegaard Research*, Jon Stewart (ed.), Hampshire: Ashgate, 2008, 117).

33 Kierkegaard, CA, 84.

34 On that *nihil*, Kierkegaard writes: "The Christian view takes the position that non-being is present everywhere as the nothing from which things were created, as semblance and vanity, as sin, as sensuousness removed from spirit, as the temporal forgotten by the eternal; consequently, the task is to do away with it in order to bring forth being" (CA, 83). More on the problematic of the beginning in chapter 3.

35 For Origen God does not *begin* to create at any point for that would imply a period prior to that point during which God remained idle (ἀργήσας ποτέ), *see De Principiis*, I, 4 (43).

36 "If the word had a beginning in time, what was God doing before the world began?" in *De Principiis*, III, 5, 238.

37 Origen, *De Principiis*, III, 5 (239).

38 Origen restricts the creation "at a definite time" to only "all *visible* things" (*De Principiis*, III, 5, 237, my emphasis).

39 "[W]hat is the change (κίνησις) of coming into existence?" Is not the whole argument of the PF in defense of the possibility for such change of a coming-into-existence, of a becoming? Especially since "this change . . . is from not existing to existing" (PF, 73). On the importance of the "moment," see Heidegger: "What we here designate as 'moment of vision' [Augenblick] is what was really comprehended for the first time in philosophy by Kierkegaard—a comprehending with which the *possibility* of a completely new epoch of philosophy has begun for the first time since antiquity. I say this is a possibility; for today when Kierkegaard has become fashionable, for whatever reasons, we have reached the stage where the literature about Kierkegaard, and everything connected with it, has ensured in all kind of ways that this decisive point of Kierkegaard's philosophy has not been comprehended" (*The Fundamental Concepts of Metaphysics*, translated by William McNeill and Nicholas Walker [Bloomington: Indiana University Press, 1995], 150).

40 Kierkegaard, CA, 242.

41 See the historical introduction to the PF by the editors, xvi–xvii.

42 Kierkegaard, CA, 25.

43 Kierkegaard, CA, 32.

44 Kierkegaard, CA, 36.

45 Kierkegaard, CA, 30. "Through the first sin, sin came into the world" (CA, 31); "*Sin came into the world by a sin*" (CA, 32, emphasis in the original).
46 Kierkegaard, CA, 40. "Furthermore, the difficulty with the serpent is something quite different, namely, that of regarding the temptation as coming from without. This is simply contrary to the teaching of the Bible, contrary to the well-known classical passage in James, which says that God tempts no man and is not tempted by anyone, but each person is tempted by himself" (CA, 48).
47 Kierkegaard, CA, 43 and 48.
48 Kierkegaard, PF, 11; my emphasis.
49 Kierkegaard, PF, 13.
50 Kierkegaard, PF, 51.
51 "As sinner, man is separated from God by the most chasmic qualitative abyss. In turn, of course, God is separated from man by the same chasmic qualitative abyss when he forgives sins. If by some kind of reverse adjustment the divine could be shifted over to the human, there is one way in which man could never in all eternity come to be like God: in forgiving sins" (Kierkegaard, *The Sickness unto Death*, edited and translated by Howard V. Hong and Edna H. Hong [Princeton: Princeton University Press, 1980], 122).
52 Kierkegaard, PF, 47.
53 Kierkegaard, CA, 56–60.
54 Kierkegaard, CA, 58.
55 Kierkegaard's remarks in pp. 58 and 59 (especially vis-à-vis the dangers of Pelagianism's return "from a different side") concern the error of making the Fall a normative state of man and nature—it should be stated clearly that, by denying a prelapsarian perfection we do not accept that the creation was created fallen, for otherwise salvation and redemption would have been abnormal.
56 Polycarp Sherwood, *The Earlier Ambigua of Saint Maximus the Confessor and his Refutation of Origenism* (Rome: Pontificium Institutum S. Anselmi, 1955), 92.
57 Maximus, *Ambiguum* 7, PG 91:1069B and 1072C.
58 Maximus the Confessor, *Ambiguum* 7, 1073C; translation by Nicholas Constas in *On Difficulties in the Church Fathers: The Ambigua, volume I*. Dumbarton Oaks Medieval Library 28. Cambridge: Harvard University Press, 2014, 87 (emphasis in the original).
59 Hans Urs von Balthasar, *Cosmic Liturgy: The Universe According to Maximus the Confessor*, translated by Brian E. Daley, S.J. (San Francisco: Ignatius, 1988), 130.

Chapter 2

1 "Video igitur tempus quondam esse distentionem." *Confessions*, XI.30.23. "Inde mihi visum est nihil esse aliud tempus quam distentionem: sed cuius rei, nescio, et mirum, si non ipsius animi" XI.33.26.
2 See Derrida, *Le Toucher* (Paris: Edition Galilée, 2000), 21ff.
3 The terminology is borrowed from Robert Sokolowski as he warns us that "[t]here is always a danger that we will separate the inseparable, that we will make the abstractum into a concretum . . ." (in *Introduction to Phenomenology* [Cambridge: Cambridge University Press, 2000], 25. William James, in a proto-phenomenological essay entitled "Does 'Consciousness' Exist?," had similarly denied the entitative aspect of consciousness (see William James, *Writings 1902–1910* [New York: The Library of America, 1987], 1141–58.

4 This is the thrust of Sartre's argument in his essay *The Transcendence of the Ego*, translated and annotated by Forrest Williams and Robert Kirkpatrick (New York: Hill and Wang, 1960).
5 "My body is thus inescapable for me: both as a condition of consciousness and as perpetually present to consciousness" in Jean-Yves Lacoste, *Experience and the Absolute: Disputed Questions on the Humanity of* Man, translated by Mark Raftery-Skehan (New York: Fordham University Press, 2004), 8. What we see around us, going by us, and carrying on with their businesses are persons not consciousness. We interact with persons not with consciousness.
6 *The Confessions* (X.16.25), 253.
7 "The analysis of time-consciousness is an ancient burden for descriptive psychology and epistemology. The first person who sensed profoundly the enormous difficulties inherent in this analysis, and who struggled with them almost to despair, was Augustine. Even today, anyone occupied with the problem of time must still study Chapters 14–28 of Book XI of the *Confessiones* thoroughly. For in these matters our modern age, so proud of its knowledge, has failed to surpass or even to match the splendid achievement of this great thinker who grappled so earnestly with the problem of time" Edmund Husserl, *On the Phenomenology of the Consciousness of Internal Time*, translated by John Barnett Brough (Boston: Kluwer Academic Publishers, 1991), 3.
8 *The Confessions* (XI.28.38), 309.
9 "It must be possible for the 'I think' to accompany all my representations" Immanuel Kant, *Critique of Pure Reason*, translated by Norman Kemp Smith (New York: Macmillan & Co., 1965), B 131, 152.
10 These paradoxes arise from our failure to recognize that multiplicity is itself multiple. Bergson distinguishes between two multiplicities that can be roughly called quantitative and qualitative. The former is derivative and secondary, a mere projection "in space" of the experience of multiplicity that the consciousness enjoys in time: "It follows from this analysis that space alone is homogeneous, that objects in space form a discrete multiplicity, and that every discrete multiplicity is got by a process of unfolding in space. It also follows that there is neither duration nor even succession in space, if we give to these words the meaning in which consciousness takes them: each of the so-called successive states of the external world exists alone; their multiplicity is real only for a consciousness that can first retain them and then set them side by side by externalizing them in relation to one another. If it retains them, it is because these distinct states of the external world give rise to states of consciousness which permeate one another, imperceptibly organize themselves into a whole, and bind the past to the present by this very process of connexion. If it externalizes them in relation to one another, the reason is that, thinking of their radical distinctness (the one having ceased to be when the other appears on the scene), it perceives them under the form of a discrete multiplicity, which amounts to setting them out in line, in the space in which each of them existed separately" in *Time and Free Will*, translated by F.L. Pogson (New York: Cosimo Classics, 1910), 120–1.
11 Husserl, *On the Phenomenology of the Consciousness of Internal Time*, translation by John Barnett Brough (Dordrecht: Kluwer, 1991), 19.
12 Husserl, *Ideas Pertaining to a Pure Phenomenology*, First Book, translated by F. Kersten (Kluwer, 1983), 94.
13 Ibid., 191.

14 This was also St. Augustine's description: "There are three tenses or times: the present of past things, the present of present things, and the present of future things" (*The Confessions*, XI.20.26, 300).

15 Edmund Husserl, *On the Phenomenology of the Consciousness of Internal Time*: the now "is continuously mediated with [the not-now], p. 42; the familiar terminology of retention and protention is introduced in p. 89; this is later called "the temporal fringe," p. 172, horizon or halo surrounding the now. See also Neal DeRoo's analysis in *Futurity in Phenomenology* (New York: Fordham University Press, 2013). This perichoretic intertwining of three "ecstasies" of time is what Heidegger calls "the horizon of time" (see, for example, *The Fundamental Concepts of Metaphysics: World, Finitude, Solitude*, translated by William McNeil and Nicholas Walker (Bloomington: Indiana University Press, 1995), especially 146–50.

16 Henri Bergson, *Time and Free Will.*

17 Ibid., 168.

18 Emmanuel Levinas, *Existence and Existents*, translated by Alphonso Lingis (The Hague: Martinus Nijhoff, 1978), 30. That fatigue is the fatigue of one's own existence is made clear in a series of remarks: "There exists a weariness which is a weariness of everything and everyone, and above all a weariness of oneself. What wearies then is not a particular form of our life—our surroundings, because they are dull and ordinary, our circle of friends, because they are vulgar and cruel; the weariness concerns existence itself" (p. 24). "In weariness we want to escape existence itself . . . Prior to every judgment, to be tired of everything and everyone is to abdicate from existence. The refusal is in weariness. Weariness by all its being effects this refusal to exist; it is only in the refusal to exist" (25).

19 Martin Heidegger, *Contributions to Philosophy*, translated by Parvis Emad and Kenneth Maly (Bloomington: Indiana University Press, 1999), 109 (§76); my emphasis. At the end of this quotation, Heidegger makes an explicit reference to his 1929/30 lecture course, published under the title *The Fundamental Concepts of Metaphysics: World, Finitude, Solitude*, which includes Heidegger's most systematic and extensive analysis of boredom. Misled by the German tern *Langeweile*, Heidegger thinks of boredom in terms of *extensity*, that is, as time that grows longer, instead of *intensity*, that it, as time that is felt "heavier." Thus, he writes: "In boredom, *Langeweile*, the while [Weile] becomes *long* [lang]" (*The Fundamental Concepts*, 152.)

20 Origen, *De Principiis*, II, 8 (English translation of Koetschau's text by G.W. Butterworth [Gloucester, Peter Smith, 1973], 125). It was on account of boredom that the souls fell away from God, a falling described as a process of losing their old fervent love, thus growing cold [*psychesthai*], from which they derived their names [*psyche*] (II, 8, 123–4).

21 And that is in spite of the few redeeming references to Kierkegaard's concept of the (kairological) moment (see *The Fundamental Concepts of Metaphysics*, 150–1).

22 Martin Heidegger, "What is Metaphysics?" translated by David Farrell Krell in *Pathmarks*, William McNeill eds, Cambridge: Cambridge University Press, 1998), 87 and *The Fundamental Concepts of Metaphysics*, 149.

23 Emmanuel Levinas, *Existence and Existents*, 435.

24 Ibid., 36.

25 As Kant has already implied (see *Critique of Pure Reason*, Transcendental Aesthetic, §6, A33/B50).

26 "How can the subject be given a definition that somehow lies in its passivity?" in Levinas, *Time and the Other*, 81.

27 Jean-Paul Sartre, *The Transcendence of the Ego*, 99–101.
28 Emmanuel Levinas, *Existence and Existents*, 75.
29 "The theory of continuous creation in Descartes and Malebranche refers, on the phenomenal level, to the incapacity of an instant to join up itself with the following instant. (. . .) Malebranche places it in its inability to preserve itself in existence in its need to resort to divine efficacity at each instant." *Existence and Existents*, 75.
30 Emmanuel Levinas, *Time and the Other*, 119. For a further reading on the concept of deformalization and time in general in Levinas's work, see Eric Severson's recent study *Levinas's Philosophy of Time: Gift, Responsibility, Diachrony, Hope* (Pittsburg: Duquesne University Press, 2013).
31 Peter Manchester in his recent book *Temporality and Trinity* (New York: Fordham University Press, 2015) treats quite insightfully the connection between Augustine's Trinitarian theology and the phenomenological analysis of temporality (in particular in Heidegger's *Being and Time*). However, I was not able to find a similar point as the one we make here that would recognize a Trinitarian trace in the very structure of the horizon of time.
32 See Hans Urs von Balthasar, *Theo-Drama V: The Last Act* (1998): "Nevertheless this fundamental quality of creaturehood (its unlikeness to God) must have some basis in God himself if it is to be posited at all" (76). And again: "A second conclusion refers to the dynamic relationship between creature and Creator, or the way in which *becoming* is rooted in absolute *Being*. We cannot avoid using the concept 'process', 'procession' in the context of the life of the Trinity to denote its constant vitality; this concept is the link between creature and Creator, between being and becoming. The eternal life that God is, and that remains 'ineffable', cannot be described as a becoming, for it is unacquainted with that 'poverty which is the ground' of our 'striving', of our 'restlessness'. 'The divine life', precisely because it is 'the fullness of life . . . [is] perfect peace'. Yet this peace, or rest, is not inert, but 'eternal movement'" (76). And finally: "Unless we see eternal *being* in terms of eternal *event*, we are condemned to see the form of its duration as a mere *nunc stans*, which deprives it of everything that makes world-time (in all its transience) exciting and delightful" (91). Emphasis in the original.
33 "The flowing-static present," cf., Husserl, *The Crisis of European Sciences*, translated by David Carr (Evanston: Northwestern University Press, 1970), 168.
34 "But we should seriously consider whether we must assume such an ultimate consciousness, which would necessarily be an 'unconscious' consciousness . . ." *On the Phenomenology of the Consciousness of Internal Time*, 394.
35 Maximus the Confessor, *Ad Thalassium* II (CCSG) 22, 59.122–59 and *Opuscula* (PG 91: 185 A). See Paul C. Plass, "'Moving Rest' in Maximus the Confessor" in *Classica et Mediaevalia*, 34 (1984), 177–90 and Sotiris Mitralexis, *Ever-moving Repose* (Berlin, 2014).
36 Basil the Great, "On Psalm 1" in *Homilies on the Psalms* (PG 29:220). Translation by Agnes Clare Way in *Exegetic Homilies*, vol. 46 (Catholic University of America Press, 2003), 158–9.
37 Ryan Coyne, *Heidegger's Confessions: The Remains of Saint Augustine in* Being and Time *& Beyond* (Chicago and London: The University of Chicago Press, 2015), 161.
38 René J. Muller, "Neurotheology: Are we Hardwired for God?" in *Psychiatric Times* (May 1, 2008).
39 Jean-Luc Marion, *Reduction and Givenness*, translated by Thomas A. Carlson (Evanston: Northwestern University Press, 1998), 197–8.

40 It is very interesting that a theologian like John Zizioulas would write of the call in terms that echo Marion's analysis that "The drive of the human being towards otherness is rooted in the divine *call* to Adam . . ." and go on to characterize the call as "the constitutive event of humanity" (see *Communion and Otherness*, 41). Nevertheless, Zizioulas's harsh criticism of consciousness as ego-centric and, consequently, of phenomenology as confined in its Cartesian heritage fails to appreciate the radically heteronomic understanding of consciousness as expressly articulated in Marion's work.
41 Jean-Luc Marion, *Being Given*, translated by Jeffrey L. Kosky (Stanford: Stanford University Press, 2002), 268.
42 For more on inverse intentionality, see my *God After Metaphysics* (Indiana University Press, 2007).
43 Marion, *Being Given*, 268.
44 Ibid.
45 Ibid., 292.
46 Jean-Luc Marion, *Idol and the Distance*, translated by Thomas A. Carlson (New York: Fordham University Press, 2001), 142.
47 Immanuel Kant, *Critique of Judgment*, translated by Werner S. Pluhar (Indianapolis: Hackett Publishing Company, 1987), 62, 64, and 221.
48 Joseph Pieper, *Divine Madness*, translated by Lothar Krauth (San Francisco: Ignatius, 1995), 48.
49 Dionysius, *On the Divine Names*, IV 7, 701C.
50 Husserl, *Ideas*, 94.
51 Ibid., 191.
52 See, for example, two of Rudolf Bernet's conclusions in his "Perception as Teleological Cognition": "Adequate givenness of a thing is not something 'real,' but rather a *teleologically anticipated idea*. (. . .) The teleological anticipation of the idea of the thing-in-itself essentially structures every perceptual process; in other words, this teleologically anticipated ideas has a *regulative function*." In *The Teleologies in Husserlian Phenomenology, Analecta Husserliana IX*, edited by Anna-Teresa Tymieniecka (Dordrecht: Reidel, 1976), 129; emphasis in the original.
53 Martin Heidegger, *History of the Concept of Time: Prolegomena*, translated by Theodore Kisiel (Bloomington: Indiana University Press, 1985), 66–7.
54 Marion, *Reduction and Givenness*, 14.
55 Augustine, *De Vera Religione* in *Augustine: Earlier Writings*, edited by J.H.S. Burleigh (Philadelphia: The Westminster Press, 1953), 253.
56 Husserl, *Ideas*, 179. "The teleological function of intentionality [in Husserl] is nothing other than this bestowal of sense." Bernard P. Dauenhauer, "The Teleology of Consciousness: Husserl and Merleau-Ponty" in *The Teleologies in Husserlian Phenomenology* (Dordrecht: Reidel, 1976), 149.
57 Augustine, *De Vera Religione*, 205.
58 Maurice Merleau-Ponty, *Le monde sensible et le monde de l'expression: Cours au Collège de France, Notes 1953* (Paris: Métis Presses, 2011), 205. See also Dauenhauer's "The Teleology of Consciousness: Husserl and Merleau-Ponty" in *The Teleologies in Husserlian Phenomenology*, 149–68.
59 Kierkegaard, CA, 58.
60 F. von G. W. Leibniz, *Theodicy*, translated by E.M. Huggard (BiblioBazaar, 2007), 145–6.
61 "But, to think time starting from the present constitutes the function, stake, and characteristic not of a specific metaphysic, but of metaphysics as a whole, from

Aristotle to Hegel (and Nietzsche)—if at least one admits the initial thought of Heidegger, hence first if one accedes to it." So Jean-Luc Marion in *God Without Being* (translation by Thomas A. Carlson, Chicago: The University of Chicago Press, 1991), 170. Hence the difference between "metaphysical" and "Christic" temporality.

62 Husserl, *Ideas*, §86.

63 Kant, *Critique of Judgment*, 68 and 31.

64 The sixth prayer in the service of the vespers, in use in the Eastern Church, speaks of God as the one who "has given to us the worldly goods as gifts, and pledged to us the promised kingdom through those goods already bestowed on us" (ὁ καὶ τὰ ἐγκόσμια ἀγαθὰ ἡμῖν δωρησάμενος καὶ κατεγγυήσας ἡμῖν τὴν ἐπηγγελμένην βασιλείαν διὰ τῶν ἤδη κεχαρισμένων ἡμῖν ἀγαθῶν).

65 Leibniz, *Theodicy*, translated by E.M. Huggard (BiblioBazaar, 2007), 45–6.

66 "Omnia appetendo proprias perfections appetunt ipsum Deum." Thomas Aquinas, *Summa Theologiae*, I., q. 6, a. I ad 2, edited and translated by Timothy McDermott, O.P. (Cambridge: Cambridge University Press, 2006), 85.

67 Husserl, *Cartesian Meditations*, translated by Dorion Cairns (The Hague: Martinus Nijhoff, 1970), §22.

Chapter 3

1 Jean-Luc Marion, *In the Self's Place: The Approach of Saint Augustine*, translated by Jeffrey L. Kosky (Stanford: Stanford University Press, 2012), 12. Hereafter abbreviated as SP.

2 See, for example, Paul Ricoeur, *Time and Narrative* (Chicago: The University of Chicago Press, 1984), vol. 1, 78.

3 SP, 13.

4 SP, 43.

5 See also the prodigal son asking for the father's fatherhood, and Oedipus becoming his own father, etc.

6 Aristotle, *Nicomachean Ethics*, IX 4, 1166a31.

7 On autonomy's denial of the origin, see William Desmond, *Ethics and the Between* (New York: SUNY Press, 2001), especially 27–38.

8 This idea becomes subject of St. Thomas's discussion on sin, especially as sin "causing" sin in *Summa Theologica*, Q. 84, 2nd Article.

9 Evelyn Waugh, *Brideshead Revisited* (Boston: Back Bay Books, 1999), 350.

10 Even our prayer is God's gift, Augustine asserts in the concluding section of the *De Dono Perseverantiae*: "Let them, therefore, notice how mistaken they are who think that we have from ourselves and are not given our asking, our seeking, and our knocking" (520).

11 That this warning was directed by St. Paul against the newly appearing Gnostics is attested by Irenaeus's interpretation who choses it as the opening of his great work *Against the Heresies* (2004).

12 SP, 222.

13 Ibid.

14 Ibid., 223.

15 Ibid.

16 Certainly, an Augustinian theme (see, for example, *The Confessions*, IV.4.9) but not without its echoes in phenomenological literature ("The *me*, as such, remains

unknown to us" would write Sartre in his early *Transcendence of the Ego* [Forrest Williams and Robert Kirkpatrick, trans., New York: Hill and Wang, 1960], 86).

17 Ibid., in particular 13–22.

18 Martin Heidegger, *Being and Time*, Division One, III ("The Worldliness of the World") and *The Fundamental Concepts of Metaphysics: World, Finitude, Solitude*, translated by William McNeill and Nicholas Walker (Bloomington: Indiana University Press, 1995), 146.

19 Martin Heidegger, *Being and Time*, Division One, II ("Being-in-the-World in General as the Fundamental Constitution of Da-Sein").

20 In a fundamental sense nature itself is unnatural: Plato's *to daimonion*, Heidegger's *das Ungeheure*, Levinas's *il y a*.

21 The foregoing distinction between space and place is indebted to Lacoste's paragraphs 1 ("Place") and 2 ("World") from his *Experience and the Absolute*, translated by Mark Reftery-Skeban (New York: Fordham University Press, 2004), 7–10 [henceforth abbreviated as EA]—which, in turn, relies upon Heidegger's analysis of the phenomenon of "being-in-the-world" (*Being and Time*, Division One, III, C, §§22–4). On the world, Lacoste writes: "I am not in the world as water is in a glass or a fish is in water. (. . .) Hence the fact that there exists no analogous relation between the world and the aquarium, in which the fish constantly comes up against its inner walls" (EA, 10).

22 Martin Heidegger, *Being and Time*, translated by Joan Stambaugh [Albany: SUNY Press, 1996], 86; Heidegger's emphasis.

23 For more on this obscure subject, see Clemens Scholten's introduction in the first volume to his (German) translation of Philoponus's *De Aeternitate Mundi* in five volumes in the series *Fontes Christiani* (Turnhout: Brepols, 2009).

24 See, for instance, Aristotle *Physics* B, 192b. Heidegger reads the Aristotelian distinction between *physis* and *techne* as "[t]wo kinds of generation" that "are contrasted with each other" ("On the Essence and Concept of Φύσις in Aristotle's Physics B, I" translated by Thomas Sheehan in *Pathmarks*, edited by William McNeill [Cambridge: Cambridge University Press, 1998], 220). Heidegger continues: "When Aristotle time and again characterizes growing things by way of analogy with artifacts, does this mean he already understands the φύσει ὄντα as self-making artifacts? No, quite the contrary, he conceives of φύσις as self-production" (p. 221). Ultimately, the natural emerges as that which has both its origin and motion from its self: the autarchic.

25 *Ktisis* (κτίσις) is the term of preference for the New Testament as well as the Patristic literature for creation, especially when contrasted with the more philosophical term *demiourgia* (δημιουργία)—even though the latter is used too by Christian sources. That creation is not to be identified with the world, see Lacoste (EA, 158): "A reality entirely different from creation, the world . . ."

26 This story is told in relation to Lacan's analysis of the *creatio ex nihilo* in *The Ethics of Psychoanalysis: 1959–1960* (The Seminar of Jacques Lacan, Book VII), edited by Jacques-Alain Miller, translated by Dennis Porter (New York and London: Norton, 1992), 122.

27 On Heraclitus famous fragment φύσις κρύπτεσθαι φιλεῖ (Diels-Kranz, fr. 123) see Pierre Hadot's *The Veil of Isis: An Essay on the History of the Idea of Nature*, translated by Michael Chase (Cambridge: Harvard University Press, 2006).

28 Dietrich Bonhoeffer, *Creation and Fall* (DBWE 3), 31–2.

29 Ibid., 32.

30 John D. Zizioulas, *The Eucharistic Communion and the World*, Luke Ben Tallon (ed.) (New York: Continuum, 2011), 158. For Zizioulas too one could "locate the heart of the problem and the crucial difference [between Christianity and Greek philosophy] in the question of whether the world has had a *beginning* or not" (157, emphasis in the original).

31 F.W.J. Schelling, *Philosophical Investigation into the Essence of Human Freedom*, translated by Jeff Love and Johannes Schmidt (Albany: SUNY Press, 2006), 28. With good reason, Schelling avoids the term "creation" for the term "procession" (*Folge*) that further confuses the life of God *ad intra* as Trinity with his creative work *ad extra*.

32 Ibid., 65. Perhaps in his attempt to affirm human freedom, Schelling undermined God's.

33 S. Kierkegaard, PF, 75.

34 Ibid., 74–5.

35 "Higher than actuality stands *possibility*" in *Being and Time*, 34. Contrary to the usual stereotypes, St. Thomas Aquinas would have agreed: "when we consider things coming to exist, potential existence precedes actual existence . . ." *Summa Theologiae*, Ia. 3, 1 (translated by Timothy McDermott O.P. [Cambridge: Cambridge University Press, 2006] vol. 2, 21. It is only "absolutely speaking" (i.e., from metaphysics' point of view) that actuality precedes possibility.

36 On contingency, see Lacoste's paragraph "Contingency and Manifestation" (although "Contingency of Manifestation" might have been more accurate according to my reading) in EA, 106–7.

37 Jean-Yves Lacoste, "The Phenomenality of Anticipation" in *Phenomenology and Eschatology: Not Yet in the Now*, Neal DeRoo and Manoussakis (eds) (Surrey: Ashgate, 2009), 19.

38 Dietrich Bonhoeffer, *Creation and Fall* (DBWE 3), 31.

39 Augustine, *The Confessions*, X, 27.38, 262.

40 Stanislas Breton in dialogue with Richard Kearney in *States of Mind: Dialogues with Contemporary Thinkers* (New York: New York University Press, 1995), 256; my emphasis.

41 Bernard P. Dauenhauer, "The Teleology of Consciousness: Husserl and Merleau-Ponty" in *The Teleologies in Husserlian Phenomenology*, 156.

42 That is Lacoste's point stated in a number of passages: "The atheism of being-in-the-world and the paganism of earthly existence correspond here to the *evident* atheism of life" (EA, 105, my emphasis). "[A]theism and paganism are inscribed in the initial structures of existence" (EA, 108). But there is more: "What is religion, if the oscillation between an existential atheism and an existential paganism . . .?" (EA, 101). In that last point, Lacoste is in agreement with Bonhoeffer's critique of "religion" and his efforts for a "religionless Christianity." A similar point is made by Jason N. Blum when writing "[i]t should be noted that both atheism and theism are *theological* positions, and neither position is scientifically established. Strictly speaking, neither explicitly theistic nor atheistic approaches to the study of religion are necessarily more defensible in metaphysical terms. This is not to say that reasons cannot be given in favor of one or the other position. (. . .) Insofar as both theism and atheism are unproven (and, as yet, unprovable) *theological* positions, however, they are on equal footing in terms of their defensibility as metaphysical assumption" (from "Retrieving Phenomenology of Religion as a Method for Religious Studies" in the *Journal of the American Academy of Religion*, December 2012, vol. 80, no. 4, 1039–40, emphasis in the original).

43 D. Bonhoeffer, *Christ the Center*, translated by Edwin H. Robertson (San Francisco: HarperCollins, 1978), 54. On the impossibility of self-autarchy one could invoke Lacoste' witness: "By working or producing, it will be said, it is also myself that I produce. It is no longer in a derived or metaphorical sense that man is said to be a creator. On the contrary, he is the creator par excellence. He is responsible for his own becoming. He does not, of course, provide himself with existence. But he contributes to his own genesis: he no doubt provides himself with the modes of being by which he concretely exists and, perhaps, also with those by which he will exist definitively. The argument is too well known for it to be necessary to develop it here" (EA, 159). It is precisely such argument that erases the concept of creation (and, subsequently, of the Creator).

44 S. Kierkegaard, *Either/Or: Part I*, edited and translated by Howard V. Hong and Edna H. Hong (Princeton: Princeton University Press, 1987), 68. And he continues: "Music does not exist except in the moment it is performed, for even if a person can read notes ever so well and has an ever so vivid imagination, he still cannot deny that only in a figurative sense does music exist when it is being read. It actually exists only when it is being performed" (ibid.).

45 This is the date of the work's first performance. In re-constructing this experience, we are aided by Thomas Forrest Kelly's informative account in *First Nights: Five Musical Premieres* (New Haven and London: Yale University Press, 2000), 108–79.

46 Nowadays the chorus is placed behind the orchestra, there is, however, speculation that at the premiere of the work the chorus was arranged before the orchestra (see Kelly, *First Nights*, 154).

47 Kelly, *First Nights*, 114.

48 Augustine, *The Confessions* (XI.2.3), 286.

49 Indeed, a number of early English translations, and even some of the twentieth century, included only the first nine books of the *Confessions* (for a review of the historical evidence, see Mark Vessey, "Book Review, *Saint Augustine: Confessions*, in *Augustinian Studies* 24, 1994, 163–81).

Chapter 4

1 Paul Ricoeur, *The Symbolism of Evil*, translated by Emerson Buchanan (Boston: Beacon Press, 1967), 203.

2 See, for example, Peter C. Bouteneff's *Beginnings: Ancient Christian Readings of the Biblical Creation Narratives* (Grand Rapids: Baker Academic, 2008).

3 James J. O'Donnell, *Augustine Confessions*, vol. II (Oxford: Oxford University Press, 2012), 126–7.

4 That the Gnostic theory of evil can be summarized by its exteriority, see Paul Ricoeur, "'Original Sin': A Study in Meaning" in *The Conflict of Interpretations*, edited by Don Ihde (Evanston: Northwestern University Press, 1974), 272.

5 Aristotle, *Nicomachean Ethics*, translated by W.D. Ross (World Library Classics, 2009), 5.

6 "...[T]he sin of my infancy (for sin there was: no one is free from sin in your sight, not even an infant whose span of earthly life is but a single day)" and a little later: "[t]he only innocent feature in babies is the weakness of their frames; the minds of infants are far from innocent" (I.7.11).

7 "Indeed, people's sins after the expulsion from paradise make Adam and Eve's partaking of the tree seem a petty offense, even if their gross disobedience is indeed terrifying." So, Peter Bouteneff in *Beginnings*, 7.
8 Paul Ricoeur's insightful analysis of the decline narratives in his *Symbolism of Evil* is indispensable here.
9 Paul Ricoeur, "'Original Sin': A Study in Meaning" in *The Conflict of Interpretations*, 272. We shall find all of the ingredients of the Gnostic myth again in Origen's system.
10 After Maximus's objection against Origen; see *Ambiguum* 42 (1325D–1336B).
11 Origen, *On First Principles* (III.6.7), translated by G. W. Butterworth (Gloucester: Peter Smith, 1973), 253.
12 Thus we read about πρώτη δημιουργία (*In insc. Pss.*, GNO V, 188), πρώτη κτίση (*In cant.*, GNO VI, 458 and in *Or. Dom.*, PG 44: 1181), πρώτη κατασκευή, πρώτη ζωή and πρώτη κοσμογένεια (in *De an.*, GNO III, 115, 112, and 119, respectively), πρώτη κατασκευή and διπλὴ κατασκευή (*De hom. opif.*, PG 44: I85B and 181B, respectively). See also Morwenna Ludlow, *Universal Salvation: Eschatology in the Thought of Gregory of Nyssa and Karl Rahner* (Oxford: Oxford University Press, 2000), 46–50. More on this in chapter 9.
13 "Διαφέρει γένεσις καὶ γέννησις. Γένεσις μὲν γάρ ἐστιν ἡ ἐκ θεοῦ πρώτη πλάσις, γέννησις δὲ ἡ ἐκ καταδίκης τοῦ θανάτου διὰ τὴν παράβασιν ἐξ ἀλλήλων διαδοχήν" in *Commentariorum series 1–145 in Matthaeum* (11), edited by Klostermann, *Origenes Werke* vol. 12 (Leipzig, 1942); ΒΕΠΕΣ, 14, 237.
14 Normally, *katabolē* is translated as the "foundation" of the world, see Mt. 13:35, 25:34; Lk. 11:50; Jn. 17:24; Eph. 1:4, and so on.
15 Origen, *On First Principles* (II.8.2), 125. The passage comes from the anathemas against Origen as decreed by the fifth ecumenical council (Constantinople, 553).
16 Origen, *On First Principles* (II.8.2), 124.
17 For example, the context of the question is here, as in Origen's theory of the two creations, Christological; Maximus deliberately uses a language that echoes that of Origen (e.g., the γένεσις/γέννησις couplet, οὐσίωσις, etc.).
18 Maximus, *Ambiguum* 42, 1325D, 24–5 (my translation).
19 After Ps. 18/19:1 "the heavens declare the glory of the God and the firmament shows his handiwork." The importance of this verse for Augustine's *Confessions* was mentioned in the last chapter. Here we can begin to appreciate the reasons for which Augustine emphasizes it.
20 Ibid., 1329Cff.
21 Ibid., 1329C, 21.
22 Ibid., 1332B. In Constas's translation, *On Difficulties in the Church Fathers*, vol. 2 (Cambridge: Harvard University Press, 2014), 153. Maximus has mentioned the Manicheans by name in the line above the passage quoted. See also Andrew Louth's assessment that "Origen, who had himself argued against many of the tenets of gnosticism, found himself in fundamental agreement with the gnostics" ("Introduction" in *Maximus the Confessor* [London and New York: Routledge, 1996], 65).
23 Maximus, *Ambiguum* 7, 1069C. Constas, *On Difficulties*, vol. 1, 79.
24 Origen, *On First Principles* (I.6.2), 53.
25 Ibid. (III.6.8), 253. Tzamalikos tries desperately to defend Origen against the accusations of Platonism (see *The Concept of Time in Origen*, Peter Lang, 1991). Yet, it is exactly Origen's willingness to let the beginning coincide with the end that betrays his (neo-) Platonist allegiance. It is precisely this aspect of Origen's thought

that Maximus corrects: "In starting from rest, the Origenists manifested their fundamental affinity with Neoplatonism, which saw the whole of reality as subject to the circular sequence of rest-processions-return. Maximus is familiar with Neoplatonic thought, and picks up several of their ideas (mainly through Denys and his first editor John of Scythopolis, or so it appears), but his rejection of Origensim in the terms we have seen entails also a fundamental rebuttal of Neoplatonism, with its ideas of emanation and return" (Louth, *Maximus the Confessor*, 67).

26 Maximus, *Centuries on Theology and Economy*, 1:66. ("Ὁ δὲ τῆς ἀναστάσεως μυηθεὶς τὴν ἀπόρρητον δύναμιν, ἔγνω τὸν ἐφ' ᾧ τὰ πάντα προηγουμένως ὁ Θεὸς ὑπεστήσατο σκοπόν" in *Philokalia*, vol. II [Athens: Aster, 1984], 61–2).

27 In Euripides, *The Complete Greek Drama*, edited by Whitney J. Oates and Eugene O'Neill, Jr. translated by E. P. Coleridge (New York. Random House, 1938).

28 Augustine, *The Confessions*, III.4.7–8, 9–80. My emphasis.

29 Cf., *The Confessions* ["This eternal Reason is your Word, who is the Beginning"] in XI.8.10 (p. 291); alluding to John 8:25.

30 Translation by Alexander Nehamas and Paul Woodruff in John M. Cooper (ed.), *Plato Complete Works* (Indianapolis and Cambridge: Hackett, 1997), 486.

31 Beginning with the Greek poet Xenophanes, this line of criticism culminates in Schleiermacher's rethinking of Christianity along the principles of Enlightenment and, further still, in Freud's construction of religion as a neurosis.

32 See the work of Jean-Luc Marion, in particular *The Idol and Distance*, translated by Thomas A. Carlson (New York: Fordham University Press, 2001).

33 Thus Hans Urs von Balthasar writes: "Who can understand that this does not give rise to two truths, because there is only one truth that is itself interpretation?" in *Theo-Logic, II: Truth of God*, Adrian J. Walker (trans.) (San Francisco: Ignatius Press, 2004), 15.

34 See Heidegger, *The Question Concerning Technology*, translated by William Lovitt (New York: Harper & Row, 1977), 7 (from where the German citation in the next sentence comes as well), and also "On the Essence and Concept of Φύσις in Aristotle's *Physics* B," in *Pathmarks*, translated by Thomas Sheehan (Cambridge: Cambridge University Press, 1998), 188.

35 Sophocles, *Oedipus at Colonus*, 1225 (the phrase is given an older provenance that leads back to Homer, see *Homeri Opera*, vol. V, edited by T.W. Allen. Oxford, 288).

36 F. Nietzsche, *Philosophy in the Tragic Age of the Greeks* (translation Marianne Cowan, Regnery Publishing, 1962, 48). Nietzsche sees in Anaximander a predecessor of Schopenhauer (46) and doesn't hesitate to identify Anaximander's "indefinite" (*apeiron*) with Kant's *Ding an sich* (47). A quite different interpretation of the same fragment is offered by Heidegger in his 1946 essay "Anaximander's Saying" (in *Off the Beaten Track*, translated by Julian Young and Kenneth Haynes, Cambridge: Cambridge University Press, 2002, 242–81). Heidegger's rendering of Anaximander's fragment—now beyond recognition—reads like that "... along the line of usage; for they let order and reck belong to one another (in the surmounting) of dis-order" (280).

37 "It is the self-consummation that consumes itself and only itself, and in its own self-consumption it finds the bitter taste of its own nothings. (...) Despair here is an empty self full of itself, circling on empty in its own lacking fullness with self." William Desmond, *Ethics and the Between* (New York: SUNY Press, 2001), 289.

38 See St. Thomas Aquinas, *Summa Theologica*, vol II, Question 75, 1st Article, where sin is defined as a negation with a deficient or an accidental efficient cause. Of sin as the "cause" of sin, see Question 84. On evil, in general, St. Thomas, following Dionysius

(*De Div. Nom.*, IV), believes that is, properly speaking, without a cause, or, if a cause is to be assigned, then that cannot be but the good (see *Summa contra Gentiles*, 3.10). That God cannot be the cause of sin, see *Questiones Disputatae de Malo*, 3.1–2.

39 F. Kafka, *The Zürau Aphorisms of Franz Kafka*, translated by Michael Hofman (New York: Schocken Books, 2006), 5 (translation modified).

40 S. Kierkegaard, CA, 49.

41 Ibid.

42 "That this desire desiring itself may end up in despair . . . is imaged in the story of Don Juan . . . the life of simulated desire of desire seeks its final 'freedom from' when it wills to be 'free from' *itself*. It cannot bear its own emptiness; it would be free from itself in a self-negation that would negate all." Desmond, *Ethics and the Between*, 288 (emphasis in the original).

43 In the twentieth chapter of the fourth book of his *Against the Heresies*, St. Irenaeus employs the image of the two hands of God saying "For God did not stand in need of these [beings] . . . as if He did not possess His own hands. For with Him were always present the Word and Wisdom, the Son and the Spirit, by whom and in whom, freely and spontaneously, He made all things . . ." (IV.20.1). However, the operation of the two hands of God is not restricted only in the creation of the world, but extends much more profoundly in the unfolding of the *economy*, that is, in the salvation of man through history; thus "the Spirit indeed working, and the Son ministering, while the Father was approving, and man's salvation being accomplished" (IV.20.6). The Word's incarnation, in particular, recapitulates times: "now this is His Word, our Lord Jesus Christ, who in the last times was made a man among men, that He might join the end to the beginning . . ." (IV.20.4). This same idea is brought to a bold summation toward the end of the fourth book by writing: "By this arrangement, therefore, and these harmonies, and a sequence of this nature, man, a created and organized being, is rendered after the image and likeness of the uncreated God—the Father planning everything well and giving His commands, the Son carrying these into execution and performing the work of creation, and the Spirit nourishing and increasing [what is made], but man making progress *day by day*, and ascending toward the perfect, that is, approximating to the uncreated One" (IV.38.3, my emphasis).

44 See, for example, "*The demonic is the sudden*" (emphasis in the original), and ". . . the negation of continuity is the sudden." Kierkegaard, CA, 129.

45 See Kierkegaard's comments on Mephistopheles' jump in Bournonville's *Faust* (CA, 131–2).

46 We shall see the connection between diachronicity and intersubjectivity in chapter 8 by way of discussing Levinas's thought.

47 NIV translation. Within square brackets I have given first the original Greek text and then the Vulgate's Latin translation.

48 A paradigmatic reading of the same parable can be found in Jean-Luc Marion's *God Without Being*, 95ff.

49 Liddell-Scott, under the entry διαλαμβάνω.

50 I. Kant, *Religion within the Boundaries of Mere Reason*, translated by Allen Wood and George di Giovanni (Cambridge: Cambridge University Press, 1998), 48.

51 Desmond, *Ethics and the Between*, 31.

52 Paul Ricoeur, *The Symbolism of Evil*, 51.

53 Ibid., 52.

54 See my *God After Metaphysics* (Bloomington: Indiana University Press, 2007), in particularly chapters one and two.

55 P. Ricoeur, *The Symbolism of Evil*, 70.
56 Nietzsche, *On the Genealogy of Morals*, translated by Walter Kaufmann and RJ Holingdale (New York: Vintage Books, 1967), 63. That idea is facilitated, of course, by the fact that in German the two terms are identical.
57 Ibid., 89. The ancient practice of worshiping the ghosts of one's ancestors, especially that of the paterfamilias (see, for example, the cult of Lares in Roman religion), seems to offer some evidence in support of Nietzsche's theory.
58 Ibid., 92, emphasis in the original.
59 See, for example, Freud's *Totem and Taboo*, translated by James Strachey (New York: Norton, 1989).
60 "All interpretation and experience of conscience agree that the "voice" of conscience somehow speaks of 'guilt'." M. Heidegger, *Being and Time*, translated by Joan Stambaugh (New York: State University of New York Press, 1996), 258. "Thus we define the formal existential idea of 'guilty' as being-the-ground for a being which is determined by a not—that is, *being-the-ground of a nullity*" (261, emphasis in the original). "Primordial being guilty cannot be defined by morality because morality already presupposed it for itself" and "Being guilty constitutes the being that we call care. Da-sein stands primordially together with itself in uncanniness. Uncanniness brings this being face to face with this undisguised nullity, which belong to the possibility of its ownmost potentiality-of-being" (264).
61 Simon Critchley, "The Null Basis-Being of a Nullity, or Between Two Nothings" in *Phenomenologies of the Stranger: Between Hostility and Hospitality*, edited by Richard Kearney and Kascha Semonovitch (New York: Fordham University Press, 2011), 150.
62 This is the case particularly for the book of Hosea. Of sin as adultery, see also Ricoeur, *The Symbolism of Evil*, 57.
63 Ibid., 76.
64 Paul Ricoeur, "'Original Sin': A Study in Meaning" in *The Conflict of Interpretations*, 282.
65 Ibid., 284.
66 Kierkegaard, CA: "At every moment, the individual is both himself and the race" (28); "[Adam] is at once himself and the race" (29); "He is not essentially different from the race, for in that case there is no race at all; he is not the race, for in that case also there would be no race. He is himself and the race. Therefore that which explains Adam also explains the race and vice versa" (ibid). For Peter Manchester "Kierkegaard maintains that Adam commits sin at the same time, that is to say in the same historical 'when,' as any individual. Adam does what 'the man' does, in a history that each of us embodies in our own biography" (*Temporality and Trinity* [New York: Fordham University Press, 2015], 11). This is correct, especially if one clarifies (as Manchester does) that this "when" of sin is not, strictly speaking, chronological but it constitutes, as Kierkegaard writes, a "qualitative leap" of the timeless *moment*.
67 Paul Ricoeur, '"Original Sin': A Study in Meaning" in *The Conflict of Interpretations*, 284 (emphasis in the original).
68 Fyodor Dostoevsky, *The Brothers Karamazov*, translated by Richard Pevear and Larissa Volokhonsky (New York: Farrar, Straus and Giroux, 1990), 320.

Chapter 5

1 F.W.J. Schelling, *Philosophical Investigations into the Essence of Human Freedom*, translated by Jeff Love and Johannes Schmidt (New York: SUNY Press, 2006), 66.

2 "[T]he first capacity for an act striving against God always remains inexplicable (. . .) this first culpability is, then, precisely already evil itself, and hence reason provides no explanation of its origin." Ibid., 25.

3 Plato, *Lysis* (220e–221). I have slightly modified Stanley Lombardo's translation (*Plato Complete Works*, edited by John M. Cooper [Indianapolis and Cambridge: Hackett, 1997], 705).

4 Ibid.

5 See Aristotle, *Poetics*, 1449b 13 and 1459a 23.

6 My reading of the Sophoclean texts is greatly indebted to a book-length discussion of *Oedipus Rex* by Ilias Papagianopoulos (*Ἐπέκεινα τῆς ἀπουσίας*, Athens: Indiktos, 2005).

7 Evelyn Waugh, *Brideshead Revisited: The Sacred and Profane Memories of Captain Charles Ryder*, Boston: Back Bay Books (1991), 21. For a philosophical interpretation of this novel, see William J. Richardson's "Psychoanalysis and the God-Question" in *Thought: A Review of Culture and Idea*, 61 no. 240 (March 1986), 68–83.

8 The Augustinian themes of *Brideshead Revisited* are numerous, perhaps the most interesting allusion is found in Sebastian Flyte's final self-exile in a monastery near Tunis (the ancient city of Carthage).

9 *Anagnorēsis* (recognition) is a key device of Greek drama, for the plot of the tragedy is usually resolved once one recognizes a character as one's own sibling (e.g., Electra or Iphigenia recognizing Orestes as their brother) etc.; the particularity of *Oedipus Rex* is that Oedipus's *anagnorēsis* consists in recognizing himself as himself (his past self).

10 It is this fact alone that unites the autobiographical narration with the philosophical treatment of time in the last three books of the *Confessions*. Further on the *Confessions*, see Jean-Luc Marion, *In the Self's Place: The Approach of Saint Augustine*.

11 A classic criticism articulated by F. Nietzsche, see for example, *Twilight of the Idols*, VII, 1, and *The Will to Power*, 272.

12 "Look upon that last day always. Count no mortal happy till/he has passed the final limit of his life secure from pain" (vs. 1529–30) as translated by David Grene (D. Grene and R. Lattimore, eds, *Greek Tragedies*, vol. 1, Chicago: University of Chicago Press, 1942).

13 For the symbolic role of the Serpent in Genesis's story, see P. Ricoeur, *The Symbolism of Evil* (Emerson Buchanan, trans.), Boston: Beacon Press, 1967. As Ricoeur notes "we seek to exculpate ourselves and make ourselves appear innocent by accusing an Other," in this case the serpent, which would then "be a part of ourselves which we do not recognize; he would be the seduction of ourselves by ourselves" (256).

14 Jean-Joseph Goux, *Oedipus, Philosopher* (Stanford: Stanford University Press, 1993), 157.

15 "[Oedipus] relies on the victories won by pure reflection and on a clear consciousness of self. His failure to recognize the forces he should have controlled and integrated is thus allowed to burgeon. Instead of letting the wise element dominate the others, the irascible and the concupiscent elements, Oedipus detaches this reasonable element, gives it a kind of autonomy and independence (by self-reflection), to such an extent that the lion and multiform beast find themselves unchained, released, liberated. Patricide and incest, even involuntarily committed, are the most searing and profound expression of that liberation, itself involuntary, unpremeditated, of two nonhuman elements. When Oedipus gets angry and kills Laïus, it is the lion element that is rebelling against the head. When Oedipus manages to share the queen's bed, it is the concupiscent element that is secretly satisfied. Each of Oedipus's involuntary crimes embodies a return of a part of the Sphinx, the return of an unconsumed and

uncomforted element of the tripartite monster that represents the monstrosity of the soul itself in its cryptic profundity." Goux, *Oedipus, Philosopher*, 157.

16 Hegel, *Aesthetics*, translated by T.M. Knox (Oxford: Oxford University Press, 1975), vol. I, 361: "It is in this sense that the Sphinx in the Greek myth, which we ourselves may interpret again symbolically, appears as a monster asking a riddle. The Sphinx propounded the well-known conundrum: What is it that in the morning goes on four legs, at mid-day on two, and in the evening on three? Oedipus found the simple answer: a man, the he tumbled the Sphinx from the rock. The explanation of the symbol lies in the absolute meaning, in the spirit, just as the famous Greek inscription calls to man: Know thyself." Hegel offers these remarks at the end of the section entitled, significantly enough, "Unconscious Symbolism." It is also interesting to note that he connects Oedipus's answer with the Delphic inscription (an allusion to Apollonian, and by extension, Socratic spirit). See also Goux, 164–6.

17 Goux, *Oedipus, Philosopher*, 177.

18 The latter Oedipus is real (as opposed to the imaginary Oedipus of the first part of the drama) insofar as he can feel pain, both emotional and physical. Thus, he assumes depth as a character.

19 As we have seen, this is how Hegel reads the story of Oedipus (*Aesthetics*, vol. 1, 361).

20 Goux, *Oedipus, Philosopher*, 161.

21 "Man being man and only man, in the innermost depths of his being, in the most invisible corners of his soul, he can no longer, in anguish and terror, encounter an other-than-himself within himself, he can no longer be invaded, seized, possessed by those disturbing and hostile forces that only images of beasts can evoke." Goux, *Oedipus, Philosopher*, 156.

22 G.K. Chesterton, *The Flying Inn* (Minneola: Dover Publications, 2001), 192.

23 For the literary genealogy of the "two paths" theme as well as its philosophical untenability, see. P. Florensky, *The Pillar and Ground of the Truth*, translated by Boris Jakim (Princeton: Princeton University Press, 1997), 124–5.

24 See Henri Bergson, *Time and Free Will: An Essay on the Immediate Date of Consciousness*, translated by F.L. Pogson (New York: Cosimo Classics, 2008), especially 175–81.

25 It is necessary to emphasize that the good is not something that can be simply given and even more "at the beginning." What we have been arguing throughout chapters 3 to 5 is precisely that the good is a temporal category linked inexorably with the end (i.e., an eschatological category).

26 It is of interest to note how the three Augustinian gardens (Thagaste, Milan, Ostia) provide the very structure of Dante's three cantos (Inferno, Purgatorio, Paradiso) in the *Divine Comedy*, indicating, perhaps, that they are not so much localities as moments and, indeed, such that in order to reach the last (Paradiso) one must go, as Dante and Virgil do, through the first (Inferno). I thank Colleen Curran for bringing this point to my attention.

Chapter 6

1 Augustine thinks throughout the *Confessions* of his relation to God in terms of distance and proximity. See also "non enim pedibus aut spatiis locorum itur abs te aut reditur ad te" (18.28). This phrase illustrates the movement of the narration in the *Confessions*: moving away from God and returning to God.

2 *Diastema* (διάστημα) belongs to Gregory of Nyssa's hermeneutical understanding of language, especially in its employment with respect to God. Gregory develops this concept as a result of his polemical engagement against Eunomius who espoused a more essentialist position in linguistic theory. For Gregory, *diastema*, which could be translated as interval or distance, pertains only to creation, in particularly in its opposition to the uncreated order (God), and it becomes another term of creatureliness. For more on this, see my *God After Metaphysics*, 96–101. Marion has suggested that the term *diastema* might be a better concept to pair with Augustine's *distentio animi*: "[r]ather than as a repetition of Plotinus, the Augustinian determination should be compared with what Gregory of Nyssa meant by διάστημα τοῦ χρόνου. This would put an even greater accent on the properly spiritual dimension (not cosmic or even psychological) of the experience of time" (*In the Self's Place*, 218).

3 See my *God After Metaphysics: A Theological Aesthetics* (Bloomington: Indiana University Press, 2007), chapter 5, especially 96–101. *Diastema* is a term employed by Gregory of Nyssa to denote the distance that separates the uncreated order (i.e., God) from the creation, in particular in reference to epistemology and language.

4 So, for example, Gregory of Nyssa, *In Canticum Canticorum* (GNO VI, 183), "solely by our inability to comprehend 'what he is' he is known 'that he is'" (τὸ ἐν μόνῳ τῷ μὴ καταλαμβάνεσθαι τί ἐστιν ὅτι ἔστιν γινωσκόμενον). See also my *God After Metaphysics*, 93–5.

5 A fall that, if read literally, with the aid of *firnamentum* used in this sentence, could suggest the Neoplatonist and Origenist idea of the soul's fall from the firmament into a human body. See O'Donnell, *Augustine Confessions*, vol. II, 48.

6 The allusion to Ricoeur's major work by the same title (*Time and Narrative*) is intended. As it is not accidental that Ricoeur himself acknowledges the exemplary position that the *Confessions* hold for his theory by opening with a reading of Augustine's text and by sustaining a dialogue with it throughout the three volumes of his work.

7 Paul Ricoeur, *Time and Narrative*, vol. 1, 54.

8 For Augustine as "the first philosopher of the Will," see Hannah Arendt's testimony in her magisterial treatment of the philosophical history of the Will in the second volume of *The Life of the Mind* (New York and London: Harcourt Brace Jovanovich, 1978), 84–110. The same conclusion is reached by Richard Sorabji, after analyzing and acknowledging the contributions of Neo-Pythagoreans, Platonists, the Stoics, the Epicureans and Plotinus ("I have ascribed to Augustine the originality of bringing all the criteria [of the concept of Will] together.") In "The Concept of the Will from Plato to Maximus the Confessor" in *The Will and Human Action*, Thomas Pink and M.W.F. Stone (eds) (London and New York: Routledge, 2004), 6–28, at 22.

9 Translation slightly modified.

10 For further discussion on the theological implications of this subject, see my "The Dialectic of Communion and Otherness in St. Maximus's Understanding of the Will" in *Knowing the Purpose of Creation Through the Resurrection, Proceedings of the Symposium on St. Maximus the Confessor, Belgrade, October 18–21, 2012*, edited by Bishop Maxim (Vasiljević) (Sebastian Press, 2013), 159–81.

11 Kierkegaard, CA, 28.

12 Jean-Paul Sartre, *The Transcendence of the Ego: An Existentialist Theory of Consciousness*, translated by Forrest Williams and Robert Kirkpatrick (New York: Hill and Wang, 1960), 100.

13 See chapter 8 below.

14 Paul Ricoeur, *Freud & Philosophy: An Essay on Interpretation*, trans. by Denis Savage (New Haven and London: Yale University Press, 1970), 378: "The Cogito, too, is a presumed certitude; it too can be deceived about itself; and no one knows to what extent. The resolute certitude of the *I am* involves the unresolved question of the possible extent of self-deception. Into this fissure, into this noncoincidence between the certitude of the *I am* and the possibility of self-deception, a certain problematic of the unconscious can be introduced."

15 Jean-Paul Sartre, *The Transcendence of the* Ego, 100, emphasis in the original.

16 Ibid., 38, my emphasis.

17 Ibid., 100, emphasis in the original.

18 In Sartre's own words: "I am condemned to be free. This means that no limits to my freedom can be found except freedom itself or, if you prefer, that we are not free to cease being free. To the extent that the for-itself wishes to hide its own nothingness from itself and to incorporate the in-itself as its true mode of being, it is trying also to hide its freedom from itself" (*Being and Nothingness*, Washington Square Press, 1956, 567). The terms "for-itself" and "in-itself" are equivalent to reflective and unreflective consciousness, respectively.

19 Jean-Paul Sartre, *Transcendence of the Ego*, 99.

20 Ibid.

21 For Husserl, it is precisely this absolute character of consciousness that differentiates it from anything else. As he writes in one instance "a veritable abyss yawns between consciousness and reality. Here, an adumbrated being, not capable of ever becoming given absolutely, merely accidental and relative; there [meaning the consciousness] a necessary and absolute being . . ." *Ideas Pertaining to a Pure Phenomenology*, First Book, translated by F. Kersten (The Hague: Martinus Nijhoff, 1983), 111.

22 "Nothing can act on consciousness, be it is cause of itself." Ibid., 82.

23 "Consciousness . . . can be limited only by itself." Ibid., 39.

24 See ibid., 61–3 (the analysis of hatred, and every other state, as "a veritable passage to infinity" [63]).

25 "An interiority closed upon itself." Ibid., 84.

26 To the extent that the ego as discussed here corresponds to the super-ego in Freud's topography of the mind, we could ascribe to it a pseudo-religious function: "[a]s a substitute for a longing for the father [the ego ideal] contains the germ from which all religions have evolved" (33); and again "[i]t is from this, indeed, that the conception arises of a higher being who deals out punishment inexorably" (56) in S. Freud's *The Ego and the Id*, trans. by Joan Riviere (New York and London: W.W. Norton, 1960).

27 There are echoes of a Hegelian theme here: see the "unhappy consciousness" from the *Phenomenology of the Spirit*.

28 For Freud an internalized Other (Father). "By setting up this ego ideal, the ego has mastered the Oedipus complex and at the same time placed itself in subjection to the id. Whereas the ego is essentially the representative of the external world, of reality, the super-ego stands in contrast to it as the representative of the internal world, of the id. Conflicts between the ego and the ideal will, as we are now prepared to find, ultimately reflect the contrast between what is real and what is psychical, between the external world and the internal world" (Freud, *The Ego and the Id*, 32). I call attention to the fact that the conflict plays out along the inside/outside polarity, as this will become important later on in our discussion of language, eating, and the bodily as the distance and the attempt to abridge this distance between inside/outside.

29 H. Bergson, *Time and Free Will*, translated by F. L. Pogson (New York: Cosimo Classics, 2008), 166–7.
30 Ibid., 128 (my emphasis). So also later: "*for the convenience of language* and the promotion of social relations, we have everything to gain by not breaking through this crust [of the parasitic self] and by assuming it to give an exact outline of the form of the object which it covers" (167, emphasis added).
31 P. Ricoeur, *Freud & Philosophy*, 388–9.
32 Sartre, *Transcendence of the Ego*, 81.
33 Ibid., 78 (and 80 where Sartre speaks of "an unintelligible spontaneity").
34 Again we appeal to Paul Ricoeur's authority. "This unawareness proper to the unreflected marks a new step toward the Freudian unconscious; it means that the co-implicit or co-intended cannot completely attain to the transparence of consciousness precisely because of the texture of the act of consciousness, i.e., because of the invincible unawareness of self that characterizes intentionality in act" (*Freud & Philosophy*, 379).
35 NIV translation, substantially revised.
36 The turning point marked by Augustine's answers to Ambrose's successor, Simplicianus (*Miscellany of Questions in Response to Simplician I*, in *Selected Writing on Grace and Pelagianism*, translated by Roland Teske, S.J. (New York: New City Press, 2011, 33–69).
37 One should follow this idea all the way to its fullest articulation by Kierkegaard who, delighting as always in a good paradox, never tired of affirming in CA that "the first sin is the sin" (30), and "[t]hrough the first sin, sin came into the world" (31), and that "sin presupposes itself" (32).
38 With reference to our need of the "daily bread," St. Maximus speaks of "the violence of nature" (διὰ τὴν βίαν τῆς φύσεως), *Expositio Orationis Dominicae*, PG 90, 904A.
39 Augustine, *The Confessions*, VIII.5.12, translated by Maria Boulding, O.S.B. (New York: New City Press, 1997), 194. Pauline terminology suggesting a rebellion of bodily members and "the law of sin in my members" are particularly apt here, given Augustine's struggle against habits formed by a licentious sexuality.
40 Ibid., VIII.5.10, p. 193.
41 The term "gnomic will" from the Greek term for deliberation (*gnōmē)* denotes the will of the willing subject as opposed to the faculty of the will common to all who share the same nature.
42 For the Maximian reading of Mt. 26:39. see *Opusculum* 6, PG 91, 65A–68D (English translation by Paul M. Blowers and Robert Louis Wilken in *On the Cosmic Mystery of Jesus Christ* [Crestwood: St. Vladimir's Seminary Press, 2003], 173–6), and *Opusculum* 7, PG 91, 69B–89B (English translation by Andrew Louth in *Maximus the Confessor*, 180–91), and François-Marie Léthel, *Théologie de l'agonie du Christ* (Paris: Beauchesne, 1979). Of key importance here is also the following passage from *Ambiguum* 7 (PG 91:10761B): "And this will take place because that which is within our power, I mean our free will—through which death made its entry among us, and confirmed at our expense the power of corruption—will have surrendered voluntarily and wholly to God, and perfectly subjected itself to His rule, by eliminating any wish that might contravene His will. And this is precisely why the Savior, exemplifying within Himself our condition, says to the Father: *Yet not as I will, but as thou wilt*" (translation by Nicholas Constas from *Maximos the Confessor, The Ambigua to Thomas and the Ambigua to John* [Cambridge, MA: Harvard University Press, 2013]). St. Augustine, too, saw in Christ's agony in Gethsemane the prefiguration and inclusion of all those Christians who would strive to unite the two Wills, so in *Enarrationes in Psalmos* 93.19

he writes: "How did our Lord marry two wills so that they become one in the humanity he bore? In his body, the Church, there would be some people who, after wanting to do their own will, would later follow the will of God. The Lord prefigured these people in himself. He wanted to show that though they are weak, they still belong to him, and so he represented them in advance in his own person. He sweated blood from his whole body, as a sign that the blood of martyrs would gush from his body, the Church ... He revealed the human will that was in him, but if he had continued to insist on that will, he would have seemed to display perversity of heart. If you recognize that he has had compassion on you, and is setting you free in himself, imitate the next prayer he made: 'Yet not what I will, but what you will be done, Father'" (*Expositions to the Psalms*, translation Maria Boulding, O.S.B. [New York: New City Press, 2002], 395).

43 Who man ultimately becomes (i.e., a man according to the flesh, the soul or the spirit) is the result of the "direction" toward which man chooses to move (i.e., the world, nature, or God, respectively)—"πρός τινα τούτων κινηθῇ *κατὰ γνώμην* ἐνδιαθέτως ὁ ἄνθρωπος" (*Letter 9*, PG 91:448A). Already in *Letter 2* it is said that the law of nature will be renewed γνωμικῶς (PG 91:396D). The same idea runs through the mature works of St. Maximus; so, for example, in *Ambiguum* 7 the willful surrender of the Will, after the model of Christ's prayer in Mt 26:39, is called ἐκχώρησιν γνωμικήν (PG 91:1076B).

44 This hypothesis is validated by von Balthasar's reading of Maximus's "critique of Origenism" (see his *Cosmic Liturgy: The Universe According to Maximus the Confessor*, translated by Brian E. Daley, S.J. [San Francisco: Ignatius, 1988], 127–36). "While motion, for Origen rested completely on the creature's undetermined freedom of will, and while this freedom, due to its extreme instability, was doomed to plunge the creature sooner or later into sin, motion for Maximus is fundamentally an orientation of nature, *which as such is good*" (130, my emphasis). Maximus constantly reminds us that God is the creator of our nature (see, for example, in *Letter 2*, PG 91:397B and 404B), the giver of our being, the originator of our motion (e.g., *Ambiguum* 7, PG 91:1076B), and so on, and therefore, nature, being and motion are as such good.

45 In *Relatio motionis* 5–120AB, see Polycarp Sherwood, *The Earlier Ambigua* (Rome: Pontificium Institutum S. Anselmi, 1955), 72. For examples of Maximus's references to Mani, see PG 91:28B and 40C.

46 Sherwood, *The Earlier Ambigua*, 204.

47 Arendt, *The Life of the Mind*, volume II: *Willing*, 42.

48 Von Balthasar, *Cosmic Liturgy*, 145, emphasis in the original.

49 "Will, therefore, man cease being a creature at the eschaton?" This is the question that Metropolitan John (Zizioulas) raises in a recent article "Eschatology and Existence" (in Greek), *Σύναξη*, 121 (2012), 63. We shall discuss his answer below.

50 On the Chalcedonean dialectic between created and uncreated, see Metropolitan John's (Zizioulas) *Communion and Otherness* (London and New York: T&T Clark, 2006), especially chapter 7, 250–85. In discussing the relationship of the will to nature we cannot lose sight of these two categories (i.e., the created and the uncreated), because the human nature as created does imply constraint and limitation—characteristics of which the divine nature is free—and therefore one should carefully avoid falling prey to speaking univocally of God's and man's nature, as it is often the case in the monothelitic debates.

51 In a number of passages St. Maximus insists that deification (i.e., salvation, grace) is not within our power, nor can be brought about by anything we can do. "We

experience divinization passively—we do not achieve it ourselves, because it lies beyond nature. For we have, within our nature, no power capable of receiving divinization." *Questiones ad Thalassium* 22 (PG 90, 324A), as quoted by von Balthasar, *Cosmic Liturgy*, 149. A lesson that needs to be remembered by those who are all too keen in offering in various spiritual manuals the steps ("purification-enlightenment-theosis") of a self-made deification, which is nothing else but the deception of self-idolatry.

52 I remind us again of Maximus's bold assertion that the goal of God is "to free man from both the world and nature" (*Letter 9*, PG 91: 448C).

53 *Letter 2* (PG 91, 396C), translated by Andrew Louth in *Maximus the Confessor* (London and New York: Routledge, 1996), 86–7. The same unity of the wills is implied as the effect of the Lord's passion "[He] made peace and reconciled us with the Father and each other through Himself, by not having any more the *gnome* resisting the *logos* of nature, but as [He had] the nature, so [He had] the *gnome* invariable" *Expositio Orationis Dominicae*, PG 90, 880A. It is important to notice that here St. Maximus ascribes a *gnome* to Christ, a position that later, during the monothelitic polemics, he retracted (see *Disputatio cum Pyrrho*, PG 91, 308D and further).

54 Lars Thuberg, *Microcosm and Mediator: The Theological Anthropology of Maximus the Confessor* (Chicago: Open Court, 1995), 228.

55 Kierkegaard, CA, 28.

56 The distance between existence and existents is best articulated and discussed by Emmanuel Levinas in one of his early works that bears this title (*Existence and Existents*, or as the French original better suggests "from Existence to Existents," translated by Alphonso Lingis [The Hague: Martinus Nijhoff, 1978]). It is there that Levinas employs for the first time the term *hypostasis* in order to name the personal existent in contradistinction to impersonal existence which has been known ever since as *il y a*.

57 "The being that is taken up [enhypostasized and made personal] is a burden." E. Levinas, *Existence and Existents*, 78.

58 Contemporary philosophy provides here an abundance of testimonials, from Kierkegaard's anxiety in the moment of decision (taken up again later by Heidegger in his essay "What is Metaphysics?") to Heidegger's care (*Sorge*) in *Being and Time*.

59 See in particular, *Opuscula* 6 and 7 (PG 91, 65A–89B); *Opusculum 15* (Spiritual and dogmatic tome against Heraclius's *Ekthesis*), PG 91: 153–84. See also note 19 above.

60 See John Gavin, S.J., "*They are like the angels in the heavens*": *Angelology and Anthropology in the Thought of Maximus the Confessor* (Rome, 2009).

61 *Confessions*, XII.11.12, translation by Maria Boulding, p. 318. Elsewhere, however (in *De Genesi ad Litteram*, 4.22.39), St. Augustine seems to accept the ability of angels to self-reflection, an ability that would suggest an act of consciousness. I am thankful to Matthew Clemente for bringing this point to my attention. We have seen in 4 that for Kierkegaard too angels are without time and without history: "Even if Michael had made a record of all the errands he had been sent on and performed, this is nevertheless not his history," CA, 49. For Kierkegaard the historical is the result of sexuality (sexual differentiation), and, therefore, of sin. "A perfect spirit cannot be conceived as sexually qualified. This is also in accord with the teachings of the Church about the nature of the Resurrection [alluding to Mt. 22:30], in accord with its representation of angels and in accord with the dogmatic definitions with respect to the person of Christ" (79).

62 Published in *Synaxe*, 121 (2012), 43–72 (in Greek).

63 Ibid., 46. The argument is made on the basis of a passage from St. Maximus's *Questiones et Dubia 13* (PG 90: 796BC).
64 Ibid., 69.
65 Ibid., 62–6.
66 *Ambiguum* 7 (PG 91:1073B), translation by Blowers and Wilken in *On the Cosmic Mystery of Jesus Christ*, 50).
67 *Ambiguum* 7 (PG 91:1401A); *Questiones ad Thalasium* 65 (PG 90: 760A); *Capita de Caritate*, 3, 25 (PG 90: 1024C). See *The Earlier Ambigua*, p. 194, note 24; and also Vasileios Betsakos's study *Στάσις Ἀεικίνητος* (Athens: Armos, 2006, in Greek).
68 *Opusculum 1* (PG 91:25AB, my translation).
69 *Opusculum 3* (PG 91, 43A), translation by Andrew Louth in *Maximus the Confessor*, 193.

Chapter 7

1 Jean-Luc Marion, *In the Self's Place*, 200.
2 Diachrony translates here as the asymmetry of my relation with the Other in the temporal which is this relation. See Emmanuel Levinas, *Time and the Other*, translated by Richard A. Cohen (Pittsburg: Duquesne University Press, 1987), 116.
3 Freud, *The Ego and the Id*, 24. And as he points out in another instance: "From a dynamic point of view [the ego] is weak, it has borrowed its energies from the id, and we are not entirely without insight into the methods—we might call them dodges—by which it extracts further amounts of energy from the id. One such method, for instance, is by identifying itself with actual or abandoned objects. The object-cathexes spring from the instinctual demands of the id. The ego has in the first instance to take note of them. But by identifying itself with the object it recommends itself to the id in place of the object and seeks to divert the id's libido on to itself." *New Introductory Lectures on Psycho-analysis*, 96.
4 Sigmund Freud, "On Narcissism: An Introduction" in *Collected Papers*, vol. IV, translation by Joan Riviere (London: The Hogarth Press, 1971), 30–59.
5 Ibid., 31.
6 Ibid., 47.
7 For an informative discussion of this famous episode, see James Wetzel's "The Trappings of Woe and Confession of Grief" in *A Reader's Companion to Augustine's Confessions*, Kim Paffenroth and Robert P. Kennedy (eds) (London: Westminster John Knox Press, 2003).
8 See Freud's reference to artistic play and in particular to tragedy on page 17.
9 S. Freud, *Beyond the Pleasure Principle*, The Standard Edition, translated and edited by James Strachey (New York and London: W.W. Norton, 1961), 33.
10 Ibid., 37.
11 "On Narcissism," 31. The same idea is expressed with greater clarity in *The Ego and the Id*: "[t]he transformation of object-libido into narcissistic libido which thus takes place obviously implies an abandonment of sexual aims, a desexualization—a kind of sublimation, therefore. Indeed, the question arises, and deserves careful consideration, whether this is not the universal road to sublimation, whether all sublimation does not take place through the mediation of the ego, which begins by changing sexual object-libido into narcissistic libido and then, perhaps, goes on to give it another aim" (24–5).

12 Kierkegaard, CA, 162.
13 "On Narcissism," 42, my emphasis.
14 As, according to B. Hopkins (*American Imago*, 38 [1981], 97–104), Augustine's theft of the pears is preceded by his feeling of guilt and not the other way around as one may assume. James O'Donnell, who preserves this reference, finds that "the question whether the guilt or the theft came first is intriguing" (*Augustine Confessions, Volume II, Commentary Book 1–7* [Oxford: Oxford University Press, 1992], 128. The theory is corroborated by Freud: "[i]n many criminals, especially youthful ones, it is possible to detect a very powerful sense of guilt which existed before the crime, and it is therefore not its result but its motive" (*The Ego and the Id*, 53).
15 Fyodor Dostoevsky, *The Brothers Karamazov*, 320.
16 Freud, *The Ego the Id*, 33.
17 Ibid., 35–6.
18 I wonder what is one to make of St. Augustine's tripartitions, or rather the trinities that he discovers in man (and in particular in man's mind) and how the Trinitarian theology the underlines these trinities might differentiate between the psychological tripartition of subordination (Plato's and Freud's) and the psychological tripartition of consubstantiality (Augustine's) where both community and otherness are simultaneously affirmed.
19 A paradigmatic reading of the same parable can be found in Marion's *God Without Being*, 95ff.
20 E. Levinas, *Existence and Existents*, trans. by Alphonso Lingis (The Hague: Martinus Nijhoff, 1978), 44.
21 It is of interest to see how this term is used elsewhere in the Scriptures and especially in the eschatological parables of the Gospel of St. Matthew. In the parable of the two sons, the father sends his first son to work in the vineyard but he refuses; "but later [ὕστερον] he changed his mind and went" (21:29). Here the term "later" is used together with a verb denoting repentance, or changing one's mind [μεταμεληθείς]. In the parable of the tenants that follows immediately after this one, the term is employed as an indicator of the time the father sent his own son to the vineyard (21:37)—a reference to the incarnation, and thus to the eschatological fullness of times.
22 Kafka, *The Blue Octavo Notebooks* (Exact Change, 1991), 28.
23 Augustine thinks throughout the *Confessions* of his relation to God in terms of distance and proximity. See also "non enim pedibus aut spatiis locorum itur abs te aut reditur ad te" (I.18.28). This phrase illustrates the movement of the narration in the *Confessions*: moving away from God and returning to God.
24 ". . . inasmuch as language distinguishes and interrelates presence and absence" (P. Ricoeur, *Freud & Philosophy*, 368–9). Ricoeur is thinking here of the original absence, the absence of the absent mother, which, in Freud's famous example, becomes regulated by (and, in turn, necessitates) the birth of language (fort/da). See also ibid., 385.
25 Ibid., 385.
26 Edmund Husserl, *Ideas Pertaining to a Pure Phenomenology*, 111. Robert Sokolowski, *Introduction to Phenomenology* (Cambridge: Cambridge University Press, 2000), 17–21.
27 Edmund Husserl, *On the Phenomenology of the Consciousness of Internal Time* and Maurice Merleau-Ponty, *Phenomenology of Perception*, trans. by Donald A. Landes (London: Routledge, 2012), especially 432–57.

28 Maurice Merleau-Ponty, Preface to A. Hesnard, *L'Oeuvre de Freud et son importance pour le mode modern* (1960), cited by Ricoeur in *Freud & Philosophy*, 417, note 99. The insight itself is discussed by Ricoeur at 382.
29 Paul Ricoeur, *Freud & Philosophy*, 383.
30 Kierkegaard, CA, 58.
31 After Husserl's remarks in the *Phenomenology of the Consciousness of Internal Time*, 81 and 390.
32 Paolo Virno, *Déjà Vu and the End of History*, translation by David Broder (London and New York: Verso, 2015), 9.
33 Andy Warhol, *The Philosophy of Andy Warhol* (Harcourt, 1975), 53.
34 Henri Bergson, "Memory of the Present and False Recognition," in *Key Writings* (London: Continuum, 2002), 148.
35 Ricoeur, *Freud & Philosophy*, 370.
36 Ibid., 383.
37 Hence the analogy between liturgical and sexual language as described by Marion in his *Erotic Phenomenon* (translated by Stephen E. Lewis, Chicago: The University of Chicago Press, 2007), 143–50.
38 I am indebted for this insight to Fr. Patrick Royannais, who writes: "Si vous voulez comprendre quelque chose à l'eucharistie, n'imaginez pas que la parole de Jésus répétée à chaque eucharistie est une information qui vise à dire ce qu'est le pain, vraiment son corps. Il s'agit bien davantage d'une déclaration d'amour. Lorsque Jésus dit, c'est mon corps pour vous, il dit, je vous aime, « Ayant aimé les siens qui étaient dans le monde, il les aima jusqu'au bout » (Jn 13) Le repas eucharistique est nuptial s'il s'agit d'alliance, comme l'enseigne Cana, festin des noces de l'agneau. Les amants se disent : prends, ceci est mon corps pour toi."
39 See, for example, Jacques Lacan's *The Ethics of Psychoanalysis*: ". . . the distance between the subject and *das Ding* [i.e., desire] . . . is precisely the condition of speech . . ." (69); and again: ". . . to the extent that the commandment in question preserves the distance from the Thing as founded by speech . . ." (83).
40 Lacan, *The Ethics of Psychoanalysis*, 309.
41 Freud, *New Introductory Lectures on Psycho-Analysis*, 92.

Chapter 8

1 Levinas, *Totality and Infinity*, 253.
2 Ibid.
3 Levinas, *Existence and Existents*, 47.
4 Indeed, there is a certain mastery involved and this is nothing else but the existent's will (as we have seen in the discussion of the will in chapter 6). My will is not that of my nature or my being. It is mine and I become acutely aware of it especially in contradistinction to my being, to existence. The tragedy of existence, however, lies in my failure—a failure predetermined and to some extent already committed—to achieve such a mastery over my being. Thus, I choose not to eat, to breathe, to live. Some of my desires are already imposed upon me and thus constitute my limitation insofar as they are for me compulsive. Is the subject then a master of being or rather its slave? Again here one can see beyond doubt that Levinas is right when he writes "there is a pain in Being" (23).

5 Levinas, *Existence and Existents*, 82. One could give the example of natural representation in icons where the planets, rivers, etc., are often depicted with faces, thus personified.
6 This brief lexical reference should indicate that the concept of the *il y a* is not an arbitrary invention of Levinas but an expression, in his language, of an older and recurring theme in the history of philosophy.
7 Levinas, *Existence and Existents*, 20.
8 Ibid.
9 Ibid., 61.
10 Ibid.
11 What is perhaps alarming in the scenarios of environmental apocalypse is not so much that human destruction will drive nature away, but rather that human destruction will efface the *order* through which nature is given, that is, nature's form or formal aspect, and therefore it will leave it in all its sheer materiality, which is unbearable. Environmental crisis reveals nature's own frightening face that has been concealed under human law. The Furies (Erinyes) if when were renamed Eumenides remain chthonic.
12 Ibid., 17.
13 Ibid., 18.
14 Ibid.
15 Ibid., 21.
16 Ibid.
17 Ibid.
18 Ibid., 76.
19 Ibid., 73.
20 Ibid., 71; emphasis in the original.
21 Ibid., 75.
22 Ibid., 26.
23 Ibid.
24 Ibid., 19.
25 Ibid.
26 Emmanuel Levinas, *Totality And Infinity: An Essay on Exteriority*, translation by Alphonso Lingis (Pittsburgh: Duquesne University Press, 1969), 78.
27 Ibid., 39.
28 Ibid., 41.
29 This term may be derived from ἀπόσταση (Greek from distance, in which case it would be synonymous with *diastema*) instead from ἀποστασία which is the term more commonly rendered as apostasy.
30 Ibid., 53.
31 Ibid., 42.
32 Ibid., 102.
33 Ibid., 34.
34 Ibid., 291.
35 Ibid., 22–3 (emphasis in the original).
36 Richard Kearney, "The Desire of God" in *God, Gift and Postmodernism*, edited by John Caputo (Bloomington: Indiana University Press, 1999), 115.
37 Ibid., 114.
38 Ibid., 115.
39 Levinas, *Totality and Infinity*, 52.

40 Ibid., 259.
41 Ibid., 260.
42 Ibid., 262.
43 Ibid., 264.
44 Ibid., 266.
45 Ibid., 263.
46 Ibid.
47 Plato, *Symposium*, 205e.
48 Ibid., 254.
49 Ibid., 265.
50 Ibid., 266–7.
51 Levinas does not retain artistic creation as an alternative to procreation. His reference to any artist at this point is restricted to the legendary figure of Pygmalion whose statues are said to have come to life (267).
52 Levinas bypasses all these problems in his haste to declare victory over "Parmenidean being." He thinks that in fecundity he has discovered the perfect way to disentangle philosophy from the asphyxiating embrace of "Eleatic unity" (277). Thus he writes (and I quote him at length): "Transcendence is time and goes unto the Other. But the Other is not a term: he does not stop the movement of Desire. The other that Desire desires is again Desire; transcendence transcends toward him who transcends—this is true adventure of paternity, of the transubstantiation which permits going beyond the simple renewal of the possible in the inevitable senescence of the subject. Transcendence, the for the Other, the goodness correlative of the face, founds a more profound relation: the goodness of goodness. Fecundity engendering fecundity accomplishes goodness: above and beyond the sacrifice that imposes a gift, the gift of the power of giving, the conception of the child. Here the Desire which in the first pages of this work we contrasted with need, the Desire that is not a lack, the Desire that is the independence of the separated being and its transcendence, is accomplished—not in being satisfied and in thus acknowledging that it was a need, but in transcending itself, in engendering Desire (269). Perhaps. But at what cost?
53 I have rehearsed some of the points I discuss here in a much earlier essay of mine under the title "Spelling Desire with Two Ls: Levinas and Lacan in *JPCS: Journal for the Psychoanalysis of Culture & Society*, vol. 7, no. 1 (Spring 2002), 16–23.
54 Jacques Lacan, *The Ethics of Psychoanalysis*, 294.
55 Ibid., 313.
56 Ibid., 314.
57 Ibid., 291.
58 Ibid., 294.
59 Ibid.
60 Ibid.
61 Mikkel Borch-Jacobsen, *Lacan: The Absolute Master*, translated by Douglas Brick (Stanford: Stanford University Press, 1991), 6.
62 "It is because for a certain point of view, the analyst is fully aware that he cannot know what is doing in psychoanalysis. Part of this action remains hidden even to him" (*The Ethics of Psychoanalysis*, 291).
63 Ibid., 309.
64 Levinas, *Totality and Infinity*, 79.
65 Levinas, "Ethics of the Infinity" in *States of Mind: Dialogues with Contemporary Thinkers*, edited by Richard Kearney (New York: New York University Press, 1995), 195.

66 Lacan, *The Ethics of Psychoanalysis*, 293 (my emphasis).
67 Levinas, "Ethics of the Infinity" in Richard Kearney's *States of Mind*, 195.
68 On our reading of Beckett's *Waiting for Godot*, see the Prolegomena, chapter 2, "Waiting Memory: Time as Intensity Through the Analysis of Boredom and Waiting."
69 Levinas, *Totality and Infinity*, 269. Zizioulas in his discussion of Levinas's "eschatology of otherness" is quite critical of him at this point; see *Communion and Otherness* (New York: T&T Clark, 2006), 48–50.
70 Lacan, *The Ethics of Psychoanalysis*, 309.
71 Maire Jaanus, "The Ethics of the Real in Lacan's Seminar VII," In *Literature and Psychology*, n. 43 (1997), 2–3.
72 Levinas, "Ethics of Infinity," 197.
73 Levinas would perhaps reply to these remarks by arguing that we can have a relation with God indeed, not with God himself but through the face of the Other, the widow, the orphan, the homeless, and so on. Does this Other (the widow, the orphan, etc.) "substitute" the absolute Other? And could one "replace" or "equate" one Other with another Other? And what kind of metonymy would this be? And, in doing so, don't I mediate the Other in terms of another Other? Don't I establish analogies that destroy the essential element of asymmetry? Let us notice also here that Levinas has rejected from the outset any notion of "incarnation" of God in the face of the Other: "The Other is not the incarnation of God" (*Totality and Infinity*, 79).
74 Kearney, "The Desire of God," 126.
75 Ibid., 127.
76 Edith Wyschogrod, *Emmanuel Levinas: The Problem of Ethical Metaphysics* (New York: Fordham University Press, 2000), xiii.
77 William Richardson, "Psychoanalysis and the God-Question" in *Thought: A Review of Culture and Idea* 51, 240 (1986), 81.
78 William Richardson, "'Like A Straw': Religion and Psychoanalysis" in *Eros and Eris*, edited by P.J.M. van Tongeren (The Hague: Kluwer Academic Publishers, 1992), 100.
79 Paul Moyaert, "Lacan on Neighborly Love: The Relation to the Thing in the Other Who Is My Neighbor" in *Epoché* 4, 1 (1996), 9–10.
80 Ibid., 4.
81 Ibid.
82 Ibid., 6.
83 Ibid., 7.
84 Jacques Derrida, "Faith and Knowledge: The Two Sources of 'Religion' at the Limits of Reason Alone" in *Religion*, edited by Jacques Derrida and Gianni Vattimo (Stanford: Stanford University Press, 1998), 65.
85 On a philosophical genealogy of this Kantian gesture, see my essay "The Philosopher-Priest and the Mythology of Reason" in *Analecta Hermeneutica* (special issue *Refiguring Divinity*) 4, 2012, 1–18.
86 Ibid., 41.
87 Ibid., 47.
88 Jacques Lacan, *The Ethics of Psychoanalysis*, 71–84 and "Kant with Sade" in *October*, 51 (Winter, 1989), 55–75.
89 Giorgio Agamben, *Potentialities*, edited and translated by Daniel Heller-Roazen (Stanford: Stanford University Press, 1999), 239.
90 Written and directed by Lars von Trier, *Dogville* was released in 2003.
91 See, for example, "the general suddenly stopped and somehow suddenly looked at his visitor in a different way; the whole change of view occurred in a single instant"

(p. 27); "[they] sensed that the décor was changing" (p. 167); "[a]gain, and as if in one instant, an extraordinary change came over him" (p. 231). Fyodor Dostoevsky, *The Idiot*, translated by Richard Pevear and Larissa Volokhonsky (Vintage Classics, 2001). I have explained the connection of the sudden with epiphany in chapter 3 of my *God After Metaphysics* (Bloomington: Indiana University Press, 2007).

Chapter 9

1 Søren Kierkegaard, *The Sickness Unto Death*, translated by Howard V. Hong and Edna H. Hong (Princeton: Princeton University Press, 1980), 7.
2 Ibid.
3 Ibid., 13.
4 Ibid.
5 Ibid., 13–14. And again: "The self is the conscious synthesis of infinitude and finitude that relates itself to itself, whose task is to become itself, which can be done only through the relationship to God" (29–30).
6 "This form of despair (ignorance of it) is the most common in the world; indeed, what we call the world, or, more exactly, what Christianity calls the world—paganism and the natural man in Christendom, paganism as it was historically and is (and paganism in Christendom is precisely this kind of despair) is despair but is ignorant of the fact." Ibid., 45.
7 Ibid., 33–4.
8 Ibid., 18.
9 Ibid., 33.
10 See our earlier discussion on chapter 4.
11 Ibid., 46.
12 Ibid., 19.
13 Ibid., 19.
14 "Imagine a self . . . and then imagine that it suddenly occurs to a self that it might become someone other—than itself. And yet one in despair this way, whose sole desire is this most lunatic of lunatic metamorphoses, is infatuated with the illusion that this change can be accomplished as easily as one changes clothes. The man of immediacy does not know himself, he quite literally identifies himself only by the clothes he wears, he identifies having a self by externalities." Ibid., 53.
15 "A beggar is naked in his consciousness, and this nakedness is only emphasized, is only intensified, becomes only all the more noticeable because of the splendor of the clothing given to him but not woven by him, not earned by him." Florensky, *The Pillar and the Ground of the Truth*, 172.
16 Odysseas Elytis, *Open Papers* in *Εκλογή 1935–1977*, edited by Eugenios Aranitses (Athens: Akmon, 1979), 146 (in Greek).
17 An term often used by Kierkegaard, translated here as "inclosing reserve" (63), it can also be rendered as "closed-in-ness," "inclosing of oneself," etc.; elsewhere (e.g., CA, 123) Kierkegaard explicitly connects *det Indesluttede* with "the demonic."
18 John Milton, *Paradise Lost*, bk 10, lines 743–5.
19 Kierkegaard, *The Sickness Unto Death*, 74.
20 Ibid., 21.
21 Pavel Florensky, *The Pillar and Ground of the Truth*, 186.
22 Ibid., 154.

23 Origen, *Exhortation to Martyrdom*, X.
24 Ibid., XXXVII.
25 1 Cor. 3:10–15, NKJV.
26 Origen, *Contra Celsum*, IV:XIII; *Ante-Nicene Fathers*, vol. 4, p. 502 (translation slightly modified).
27 Theodore the Studite, *Letter 52*, PG 99, 1501AB.
28 Florensky, *The Pillar and Ground of the Truth*, 173.
29 "But a creature of God is a person and must be saved. An evil character is precisely what prevents a person from being saved. It is therefore clear that salvation postulates a separation between person and character." Ibid., 156.
30 Ibid., 178. On the notion of the "second death," see *Revelation*, 2:11, 20:6, 20:14 and 21:8.
31 Lacan, *The Ethics of Psychoanalysis*, 295.
32 Florensky, *The Pillar and Ground of the Truth*, 178 (emphasis in the original).
33 Ibid., 171.
34 Ibid., 175.
35 Kierkegaard, *Journals and Papers*, XI 2 A 164 n.d. 1854 (Hong and Hong, vol. 1, 152).
36 Florensky, *The Pillar and Ground of the Truth*, 176.
37 Gregory of Nyssa, *De Anima et Resurrectione* [*An et res*] in *Gregorrii Nysseni Opera* [GNO], volume III, edited by Andreas Spira (Leiden and Boston: Brill, 2014).
38 I do not enter here in the debate whether Macrina is merely a character in Gregory's text, styled perhaps so as to voice a more Platonic version of Christianity than what Gregory himself espouses in other works of his. However, I believe that such a possibility should be taken seriously into consideration in further readings of the dialogue.
39 *An et res*, 105.
40 Absent here is the distinction between resuscitation and resurrection as found, for example, in Epiphanios of Salamis.
41 On Gregory's employment of this distinction and on the meaning of *diastema*, see Hans Urs von Balthasar's *Presence and Thought: An Essay on the Religious Philosophy of Gregory of Nyssa*, translated by Mark Sebanc (San Francisco: Ignatius, 1995), in particular "The Concept of Spacing" (27–35).
42 "γενητή": *An et res*, 15.
43 *Contra Eun.* III, GNO ii, 210 (Η δὲ θεία φύσις [. . .] ἀδιάστατος καὶ ἄποσος καὶ ἀπερίγραπτος), 217 (οὐδὲ τινος διαστηματικῆς ἐννοίας περὶ τῆν θείαν φύσιν τὴν ἄποσον τε καὶ ἀδιάστατον θεωρουμένης).
44 See, for example, τὴν νοερὰν ταύτην καὶ ἀδιάστατον φύσιν, ἣν καλοῦμεν ψυχήν (*De an.*, 29); ἡ δὲ νοερά τε καὶ ἀδιάστατος φύσις . . . οὐκοῦν ἐστιν ἐν αὐτοῖς ἡ ψυχή (*De an.*, 31).
45 See *De an.*, 54 (τῇ γνωστικῇ δυνάμει τοῦ οἰκείου [ἡ ψυχή] ἐφαπτομένη καὶ παραμένουσα); 55 (τούτοις καὶ μετὰ τὴν διάλυσιν [ἡ ψυχή] παραμένει οἱονεὶ φύλακα τῶν οἰκείων καθισταμένην). It is interesting that on the basis of this affinity between soul and body Gregory is able to sketch a theory on relics (and perhaps in defense of their veneration), see *De an.*, 57. The term *λείψανον* allows him to use the metaphor of the broken pottery and, at the same time, allude to the saint's relics. On the same subject, see John Gavin's "Souls and Bodies, History and Eternity: John Scottus Eriugena and the Intermediate State" in *Anales Theologici* 29 (2015): 139–54.
46 This is Gregory's (as one of the *personae dramatis*) longest contribution to the dialogue; it runs from page 105 to page 111.

47 “δῆμος τις ἀνθρώπων πάντως ὁ εἷς γενήσεται,” (ibid., 109).
48 “ἡ εἰς τὸ ἀρχαῖον τῆς φύσεως ἡμῶν ἀποκατάστασις,” (ibid., 112). And again: “ὅταν δὲ πρὸς τὴν πρώτην τοῦ ἀνθρώπου καταστευὴν δι᾽ἀναστάσεως ὁ θεὸς ἐπανάγῃ τὴν φύσιν,” (ibid., 115).
49 “μετασχεῖν τῆς μακαριότητος,” (ibid., 116).
50 *De an.*, 115–16, translation by William Moore and Henry Austin Wilson in NPNF 5, edited by Philip Schaff and Henry Wace (Buffalo, NY: Christian Literature Publishing Co., 1893); translation modified.
51 “καὶ μὴ λήγει ποτὲ τῆς αὐξήσεως,” (ibid., 78).
52 “ἐφ᾽ὧν ὅρος οὐδείς ἐπικόπτει τὴν αὔξησιν,” (ibid., 79). Gregory’s emphasis here is also directed against Origen’s doctrine of satiation (κόρος), which he has addressed earlier (see 71). For Origen’s position, see *De Principiis*, II, 8, 123–5).
53 Maximus, *Cap. Caritate*, 3, 25 (PG 90, 1024C). See also the distinction in Zizioulas’s «Εσχατολογία καὶ Ὕπαρξη», 62–4. This distinction between the two kinds of eternity is also found in Heidegger’s *Hölderlins Hymnen Germanien und Der Rhein*, in *GA* 39, 55.
54 *De an.*, 92 (translation modified).
55 “τῇ τῆς κακίας εἰσόδῳ εἰς πλῆθος ἡ φύσις κερματίσθη,” (ibid., 120).
56 Morwenna Ludlow writes with respect to Gregory’s notion of the first creation: “it is almost as if Gregory thinks that ideal humanity or human nature (ἡ ἀνθρώπινη φύσις) is akin *to a Platonic form in the mind of God*. It thus already in a sense exists . . .” (in *Universal Salvation: Eschatology in the Thought of Gregory of Nyssa and Karl Rahner* [Oxford: Oxford University Press, 2000], 49, my emphasis). We would like to add that not only it exists “in a sense,” but preeminently so, for to be “in God’s mind” is to exist most truly and properly. We shall avoid entering here into a discussion of Patristic exemplarism.
57 *An et res.*, 119.
58 What sense is one to make out of such a hope which is nothing more than the return to the first condition (τὸ ἐλπιζόμενον οὐδὲν ἕτερον ἐστιν ἢ ὅπερ ἐν πρώτοις ἦν, 120) other than nostalgia for an imaginary beginning?
59 *De an.*, 59.
60 Desires (conceived precisely as the movements of the soul—τὰ τῆς ψυχῆς κινήματα, 65) were altogether and indiscriminately ruled out from the eschatological condition (see 66–8).
61 Cardinal Ratzinger (Pope Benedict XVI) raises a very pertinent point with regards to the relation between the dead and history: “When we die, we step beyond history. In a preliminary fashion history is concluded—for me. But this does not mean that we lose our relation to history: the network of human relationality belongs to human nature itself. History would be deprived of its seriousness if resurrection occurred at the moment of death” (*Eschatology: Death and the Eternal Life*, translated by Michael Waldstein [Washington, D.C.: The Catholic University of America Press, 1988], 184).
62 *In cant.*, GNO vi, 457–8. See also Ludlow, *Universal Salvation*, 49.
63 *De an.*, 60 (translation from *NPNF*).
64 Ibid., 110.
65 As we have already seen above (*An et res.*, 112 and 115). On the other hand: “. . . not only will the body not be such [eschatologically] as it is now, even when it is in the best of health; *it will not even be such as it was in the first human beings prior to sin*” (St. Augustine, *De Civ. Dei*, XIII. 20, 87 my emphasis).
66 “Ταῦτα δὲ πάντα τῇ εἰσόδῳ τῆς κακίας ἡμῖν συνεισέβαλεν,” (ibid., 113).

67 Contrast St. Gregory's silence over Christ's risen body to this: "even after his [i.e., Christ's] resurrection, in his already spiritual but still real flesh, he took food and drink with his disciples. For it is not the ability but rather the need to eat and drink that will be removed from such bodies. These bodies will be spiritual, therefore, not because they will cease to be bodies but because they will be sustained by a life-giving spirit" (St. Augustine, *De Civ. Dei*, XIII. 22, 89).
68 "ὁμοιότης," (ibid., 26).
69 On this similitude between the divine nature and the human soul, see Salvatore R.C. Lila, *Neuplatonisches Gedankengut in den "Homilien über die Seligpreisungen" Gregors von Nyssa*, edited by Drobner (Leiden 2004), 61–7.

Chapter 10

1 "I confess that we shall all rise in this flesh." Gregory the Great attributes this saying as the last words of Eutychius, Patriarch of Constantinople (*Moralia*, XIV, 56).
2 W. M. Sinclair, "Eutychius" in *Dictionary of Christian Biography and Literature to the End of Sixth Century*, edited by William C. Piercy and Henry Wace (London, John Murray, 1880), 415.
3 Gregory the Great, *Moralia*, XIV, 56.
4 Gregory the Great, *Morals on the Book of Job* (XIV, 56), translated by John Henry Parker (Oxford, 1844).
5 Gregory of Nyssa, *An et res.*, GNO iii, 55.
6 In NKJV; an alternative reading of the last verse: "If there is a natural body, then there is a spiritual body [as well]."
7 See, for example, Dale Martin's *The Corinthian Body* (New Haven: Yale University Press, 1995); Jeffrey Asher, *Polarity and Change in 1 Corinthians 15* (HUT 42, Tübingen: Mohr Siebeck, 2000); and Engberg-Pedersen, *Cosmology and Self in the Apostle Paul: The Material Spirit* (Oxford: Oxford University Press, 2010).
8 James Ware, "Paul's Understanding of the Resurrection in 1 Corinthians 15:36–54," in *Journal of Biblical Literature*, 133, no. 4 (2014), 832 (emphasis in the original).
9 Emmanuel Falque, *God, the Flesh and the Other: From Irenaeus to Duns Scotus*, translated by William Christian Hackett (Evanston: Northwestern University Press, 2015), 131.
10 Ibid., emphasis in the original.
11 Ibid., 130.
12 Gregory the Great, *Morals on the Book of Job* (XIV, 56).
13 Emmanuel Falque, *God, the Flesh and the Other: From Irenaeus to Duns Scotus.*
14 Eustratius of Constantinople, *De Statu Animarum Post Mortem*, edited by Peter van Deun, *Corpus Christianorum Series Graeca* 60 (Turnhout: Brepols, 2006).
15 Methodius of Olympus, *On the Resurrection* (PG 18: 325–8), in *Ante-Nicene Fathers*, vol. 6, 377.
16 Gregory of Nyssa, *De hominis opificio* (henceforth abbreviated as *Op. hom.*), PG 44: 225.
17 Ibid.
18 Gregory of Nyssa, *De an.*, p. 57.
19 Ibid., 54. In light of these passages, we should credit Gregory of Nyssa with the honor of anticipating the idea of a *horcrux* some centuries before it became popular.

20 John Scottus Eriugena, *Periphyseon* III, edited by É. A. Jeauneau (Turnholt: Brepols, 1999), p. 163; as cited in John Gavin, "Souls and Bodies, History and Eternity: John Scottus Eriugena and the Intermediate State" in *Annales Theologici* 29 (2015), 150.
21 Ibid.
22 See Falque's gloss on 2 Cor. 4:11: ". . . for surely the Hellenizing dualism of *psuchê/sôma* is very far removed from the biblical conception of man as *sarx*." In *God, the Flesh, and the Other*, 157 (see also 149).
23 John Damascene, *De Fide Orthodoxa*, II:3, PG 84: 868B: "Ἀνεπίδεκτος μετανοίας, ὅτι καὶ ἀσώματος. Ὁ γὰρ ἄνθρωπος διὰ τὴν τοῦ σώματος ἀσθένειαν μετανοίας ἔτυχεν."
24 Freud, *New Introductory Lectures*, 91.
25 This distinction is made most clearly in *Crisis*, §28, 107, although frequently employed in his earlier works.
26 Lacan, *Seminar XI: The Four Fundamental Concepts of Psycho-Analysis*, translated by A. Sheridan (New York: Norton, 1978).
27 Further evidence in support of this connection can be gleamed from Merleau-Ponty's insight that the body in its ambiguous state (neither consciousness nor a thing of the world) is phenomenology's closest approximation to the Freudian unconscious. See his Preface to A. Hesnard, *L'Oeuvre de Freud et son importance pour le mode modern* (1960), cited by Ricoeur in *Freud & Philosophy*, 417, note 99. See also Emmanuel Falque, *God, the Flesh, and the Other: From Irenaeus to Duns Scotts*, 117 and ff.
28 Jacques Lacan, *The Ethics of Psychoanalysis*, 55. There are several other references to Kant in connection to *das Ding* in this work.
29 See Kevin Hart's introduction to saturated phenomena in *Jean-Luc Marion: Essential Writings* (New York: Fordham University Press, 2013), 76–7.
30 Jean-Luc Marion, *In Excess: Studies of Saturated Phenomena*, translated by Robyn Horner and Vincent Berraud (New York: Fordham University Press, 2002), 112.
31 Martin Heidegger, "The Thing" in *Poetry, Language, Thought*, translated by Albert Hofstadter (New York: Perennial Classics, 2001), 165–6.
32 Ibid., 167.
33 Ibid.
34 Ibid.
35 Ibid., 169.
36 Ibid., 170.
37 "It is in this way that another typology is established, the typology which institutes the relation to the real. And now we can define this relation to the real, and realize what the reality principle means." Jacques Lacan, *The Ethics of Psychoanalysis*, 66.
38 Lacan, *The Ethics of Psychoanalysis*, 139.
39 See Augustine, *Confessions*, III.11.20, 90; the reference is to an interpretation of Monica's dream.
40 Jacques Lacan, *The Ethics of Psychoanalysis*, 66: "The Other of the Other only exists as a place." And again, "I mean that the whole development at the level of the mother/child interpsychology—and that is badly expressed in the so-called categories of frustration, satisfaction, and dependence—is nothing more than an immense development of the essential character of the maternal thing, of the mother, *insofar as she occupies the place of that thing*, of *das Ding*" (67, emphasis added).
41 Paul Moyaert, "Lacan on Neighborly Love: The Relation to the Thing in the Other Who Is My Neighbor" in *Epoché* 4, 1 (1996), 7.
42 Luigi Pirandello, "The Oil Jar" in *Undici Novelle*, translated by Stanley Appelbaum (New York: Dover, 1994).

43 Luigi Pirandello, "The Oil Jar," 103.
44 Ibid., 105.
45 Ibid., 101.
46 Ibid., 99.
47 Ibid., 109.
48 Jacques Lacan, *The Ethics of Psychoanalysis*, 121.
49 Ibid., 71.
50 On this connection, see Marcus Pound, *Theology, Psychoanalysis and Trauma* (London: SCM Press, 2007).

Bibliography

Agamben, Giorgio. *Potentialities*. Translated by Daniel Heller-Roazen. Stanford: Stanford University Press, 1999.

Alberigo, G., ed. *Conciliorum Oecumenicorum Generaliumque Decreta*. Turnhout: Brepols, 2006.

Arendt, Hannah. *The Life of the Mind: Willing*. New York and London: Harcourt Brace Jovanovich, 1978.

Aristotle. *Physics*. Text with introduction by W.D. Ross. Oxford: Oxford University Press, 1936.

Aristotle. *Nicomachean Ethics*. Translated by W.D. Ross. World Library Classics, 2009.

Asher, Jeffrey. *Polarity and Change in 1 Corinthians 15*. Tübingen: Mohr Siebeck, 2000.

Augustine. Of True Religion (*De Vera Religione*). In *Augustine: Earlier Writings*, edited by J.H.S. Burleigh. Philadelphia: The Westminster Press, 1953.

Augustine. *The Confessions*. Translated by Maria Boulding. New York: New City Press, 1997.

Augustine. *Expositions to the Psalms*. Translated by Maria Boulding. New York: New City Press, 2002.

Augustine. *Confessiones*. Edited by Martinus Skutella. Berlin: De Gruyter, 2009.

Augustine. *The Gift of Perseverance* (*De Dono Perseverantiae*). In *Selected Writings on Grace and Pelagianism*, edited by Boniface Ramsey and translated by Roland Teske. New York: New City Press, 2011.

Augustine. *Miscellany of Questions in Response to Simplician I*. In *Selected Writing on Grace and Pelagianism*, translated by Roland Teske. New York: New City Press, 2011.

Augustine. *The City of God*. Translated by William Babcock. New York: New City Press, 2013.

Basil the Great. *Homilies on the Psalms*. In *Exegetic Homilies*, translated by Agnes Clare Way. Washington, DC: Catholic University of America Press, 2003.

Bergson, Henri. *Time and Free Will*. Translated by F.L. Pogson. New York: Cosimo Classics, 1910.

Bergson, Henri. "Memory of the Present and False Recognition." In *Henri Bergson: Key Writings*. Edited by John Mullarkey and Keith Ansell Pearson. London: Continuum, 2002.

Bernet, Rudolf. "Perception as Teleological Cognition." In *The Teleologies in Husserlian Phenomenology, Analecta Husserliana IX*, edited by Anna-Teresa Tymieniecka. Dordrecht: Reidel, 1976.

Betsakos, Vasileios. *Στάσις Ἀεικίνητος*. Athens: Armos, 2006.

Blum, N. "Retrieving Phenomenology of Religion as a Method for Religious Studies." In the *Journal of the American Academy of Religion*, 80:4 (December 2012): 1039–40.

Bonhoeffer, Dietrich. *Christ the Center*. Translated by Edwin H. Robertson. San Francisco: HarperCollins, 1978.

Bonhoeffer, Dietrich. *Creation and Fall*. Translated by Douglas Stephen Bax. Minneapolis: Fortress Press, 1997.

Borch-Jacobsen, Mikkel. *Lacan: The Absolute Master*. Translated by Douglas Brick. Stanford: Stanford University Press, 1991.

Bouteneff, Peter C. *Beginnings: Ancient Christian Readings of the Biblical Creation Narratives*. Grand Rapids: Baker Academic, 2008.

Carlisle, Clare. *Kierkegaard's Philosophy of Becoming: Movements and Positions*. New York: SUNY Press, 2005.

Chesterton, G.K. *The Flying Inn*. Minneola: Dover Publications, 2001.

Coyne, Ryan. *Heidegger's Confessions: The Remains of Saint Augustine in* Being and Time *& Beyond*. Chicago and London: The University of Chicago Press, 2015.

Critchley, Simon. "The Null Basis-Being of a Nullity, or Between Two Nothings." In *Phenomenologies of the Stranger: Between Hostility and Hospitality*, edited by Richard Kearney. New York: Fordham University Press, 2011.

Curd, Patricia. *Anaxagoras of Clazomenae*. Toronto: University of Toronto Press, 2010.

Dauenhauer, Bernard P. "The Teleology of Consciousness: Husserl and Merleau-Ponty." In *The Teleologies in Husserlian Phenomenology, Analecta Husserliana IX*, edited by Anna-Teresa Tymieniecka. Dordrecht: Reidel, 1976.

DeRoo, Neal. *Futurity in Phenomenology*. New York: Fordham University Press, 2013.

Derrida, Jacques. "Faith and Knowledge: The Two Sources of 'Religion' at the Limits of Reason Alone." In *Religion*, edited by Jacques Derrida and Gianni Vattimo. Stanford: Stanford University Press, 1998.

Derrida, Jacques. *Le Toucher*. Paris: Edition Galilée, 2000.

Desmond, William. *Ethics and the Between*. New York: SUNY Press, 2001.

Dionysius. *De Divinis Nominibus*. In *Corpus Dionysiacum I*, edited by Beate Regina Suchla. Berlin: De Gruyter, 1990.

Dostoevsky, Fyodor. *The Brothers Karamazov*. Translated by Richard Pevear and Larissa Volokhonsky. New York: Farrar, Straus and Giroux, 1990.

Dostoevsky, Fyodor. *The Idiot*. Translated by Richard Pevear and Larissa Volokhonsky. New York: Vintage Classics, 2001.

Elytis, Odysseas. *Open Papers*. In *Εκλογή 1935–1977*. Edited by Eugenios Aranitses. Athens: Akmon, 1979.

Engberg-Pedersen, Troels. *Cosmology and Self in the Apostle Paul: The Material Spirit*. Oxford: Oxford University Press, 2010.

Euripides. *Helen*. In *The Complete Greek Drama*, edited by Whitney J. Oates and Eugene O'Neill, Jr., translated by E. P. Coleridge. New York. Random House, 1938.

Eustratius of Constantinople. *De Statu Animarum Post Mortem*. In *Corpus Christianorum Series Graeca* 60, edited by Peter van Deun. Turnhout: Brepols, 2006.

Falque, Emmanuel. *God, the Flesh and the Other: From Irenaeus to Duns Scotus*. Translated by William Christian Hackett. Evanston: Northwestern University Press, 2015.

Florensky, Pavel. *The Pillar and Ground of the Truth*. Translated by Boris Jakim. Princeton: Princeton University Press, 1997.

Freud, Sigmund. *The Ego and the Id*. Translated by Joan Riviere. New York and London: W.W. Norton, 1960.

Freud, Sigmund. *Beyond the Pleasure Principle*. Translated and edited by James Strachey. New York and London: W.W. Norton, 1961.

Freud, Sigmund. "On Narcissism: An Introduction." In *Collected Papers*, volume IV, translated by Joan Riviere. London: The Hogarth Press, 1971.

Freud, Sigmund. *Totem and Taboo*. Translated by James Strachey. New York: Norton, 1989.

Gavin, John. "*They are like the angels in the heavens*": *Angelology and Anthropology in the Thought of Maximus the Confessor*. Rome, 2009.

Gavin, John. "Souls and Bodies, History and Eternity: John Scottus Eriugena and the Intermediate State." In *Annales Theologici* 29 (2015): 139–54.
Goux, Jean-Joseph. *Oedipus, Philosopher*. Stanford: Stanford University Press, 1993.
Gregory of Nyssa. *De Anima et Resurrectione*. In *Gregorrii Nysseni Opera*, volume III, edited by Andreas Spira. Leiden and Boston: Brill, 2014.
Gregory the Great. *Morals on the Book of Job*. Translated by John Henry Parker. Oxford, 1844.
Hadot, Pierre. *The Veil of Isis: An Essay on the History of the Idea of Nature*. Translated by Michael Chase. Cambridge: Harvard University Press, 2006.
Hart, Kevin. "Introduction." In *Jean-Luc Marion: Essential Writings*. Edited by Kevin Hart. New York: Fordham University Press, 2013.
Hegel. *Aesthetics*. Translated by T.M. Knox. Oxford: Oxford University Press, 1975.
Heidegger, Martin. *The Question Concerning Technology*. Translated by William Lovitt. New York: Harper & Row, 1977.
Heidegger, Martin. *History of the Concept of Time: Prolegomena*. Translated by Theodore Kisiel. Bloomington: Indiana University Press, 1985.
Heidegger, Martin. *The Fundamental Concepts of Metaphysics: World, Finitude, Solitude*. Translated by William McNeill and Nicholas Walker. Bloomington: Indiana University Press, 1995.
Heidegger, Martin. *Being and Time*. Translated by Joan Stambaugh. Albany: SUNY Press, 1996.
Heidegger, Martin. "On the Essence and Concept of Φύσις in Aristotle's Physics B, I." In *Pathmarks*, edited by William McNeill and translated by Thomas Sheehan. Cambridge: Cambridge University Press, 1998.
Heidegger, Martin. "What is Metaphysics?" In *Pathmarks*, edited by William McNeill and translated by David Farrell Krell. Cambridge: Cambridge University Press, 1998.
Heidegger, Martin. *Contributions to Philosophy*. Translated by Parvis Emad and Kenneth Maly. Bloomington: Indiana University Press, 1999.
Heidegger, Martin. "The Thing." In *Poetry, Language, Thought*. Translated by Albert Hofstadter. New York: Perennial Classics, 2001.
Heidegger, Martin. "Anaximander's Saying." In *Off the Beaten Track*, translated by Julian Young and Kenneth Haynes. Cambridge: Cambridge University Press, 2002.
Hopkins, Brooke. "St. Augustine's Confessions: The Pear-Stealing Episode." In *American Imago*, 38 (1981): 97–104.
Husserl, Edmund. *Cartesian Meditations*. Translated by Dorion Cairns. The Hague: Martinus Nijhoff, 1970.
Husserl, Edmund. *The Crisis of European Sciences*. Translated by David Carr. Evanston: Northwestern University Press, 1970.
Husserl, Edmund. *Ideas Pertaining to a Pure Phenomenology: First Book*. Translated by F. Kersten. The Hague: Kluwer, 1983.
Husserl, Edmund. *On the Phenomenology of the Consciousness of Internal Time*. Translated by John Barnett Brough. Boston: Kluwer Academic Publishers, 1991.
Irenaeus. *Against the Heresies*. In *Ante-Nicene Fathers*, edited by Alexander Roberts. Peabody: Hendrickson Publishers, 2004.
Jaanus, Maire. "The Ethics of the Real in Lacan's Seminar VII." In *Literature and Psychology*, 43 (1997): 1–18.
James, William. "Does 'Consciousness' Exist?" In *William James: Writings 1902–1910*. New York: The Library of America, 1987.
Kafka, Franz. *The Blue Octavo Notebooks*. Exact Change, 1991.

Kafka, Franz. *The Zürau Aphorisms of Franz Kafka*. Translated by Michael Hofman. New York: Schocken Books, 2006.

Kangas, David J. *Kierkegaard's Instant: On Beginnings*. Indianapolis: Indiana University Press, 2007.

Kant, Immanuel. *Critique of Pure Reason*. Translated by Norman Kemp Smith. New York: Macmillan, 1965.

Kant, Immanuel. *Critique of Judgment*. Translated by Werner S. Pluhar. Indianapolis: Hackett Publishing Company, 1987.

Kant, Immanuel. *Religion within the Boundaries of Mere Reason*. Translated by Allen Wood and George di Giovanni. Cambridge: Cambridge University Press, 1998.

Kearney, Richard. *States of Mind: Dialogues with Contemporary Thinkers*. New York: New York University Press, 1995.

Kearney, Richard. "The Desire of God." In *God, Gift and Postmodernism*, edited by John Caputo. Bloomington: Indiana University Press, 1999.

Kelly, Forrest Thomas. *First Nights: Five Musical Premieres*. New Haven and London: Yale University Press, 2000.

Kierkegaard, Søren. *The Concept of Anxiety*. Edited and translated by Reidar Thomte. Princeton: Princeton University Press, 1980.

Kierkegaard, Søren. *The Sickness unto Death*. Edited and translated by Howard V. Hong and Edna H. Hong, Princeton: Princeton University Press, 1980.

Kierkegaard, Søren. *Philosophical Fragments*. Edited and translated by Howard V. Hong and Edna H. Hong. Princeton, Princeton University Press, 1985.

Kierkegaard, Søren. *Either/Or: Part I*. Edited and translated by Howard V. Hong and Edna H. Hong. Princeton: Princeton University Press, 1987.

Kierkegaard, Søren. *Concluding Unscientific Postscript to the Philosophical Fragments*. Edited and translated by Howard V. Hong and Edna H. Hong. Princeton: Princeton University Press, 1992.

Lacan, Jacques. *Seminar XI: The Four Fundamental Concepts of Psycho-Analysis,*. Translated by A. Sheridan. New York: Norton, 1978.

Lacan, Jacques. "Kant with Sade." Translated by James B. Swenson. In *October*, 51 (Winter, 1989): 55–75.

Lacan, Jacques. *Seminar VII: The Ethics of Psychoanalysis*. Edited by Jacques-Alain Miller and translated by Dennis Porter. New York and London: Norton, 1992.

Lacoste, Jean-Yves. *Experience and the Absolute: Disputed Questions on the Humanity of Man*. Translated by Mark Raftery-Skehan. New York: Fordham University Press, 2004.

Lacoste, Jean-Yves. "The Phenomenality of Anticipation." In *Phenomenology and Eschatology: Not Yet in the Now*, edited by Neal DeRoo and John Panteleimon Manoussakis. Surrey: Ashgate, 2009.

Leibniz. *Theodicy*. Translated by E.M. Huggard. BiblioBazaar, 2007.

Léthel, François-Marie. *Théologie de l'agonie du Christ*. Paris: Beauchesne, 1979.

Levinas, Emmanuel. *Totality and Infinity: An Essay on Exteriority*. Translated by Alphonso Lingis. Pittsburgh: Duquesne University Press, 1969.

Levinas, Emmanuel. *Existence and Existents*. Translated by Alphonso Lingis. The Hague: Martinus Nijhoff, 1978.

Levinas, Emmanuel. *Time and the Other*. Translated by Richard A. Cohen. Pittsburg: Duquesne University Press, 1987.

Levinas, Emmanuel. "Ethics of the Infinity." In *States of Mind: Dialogues with Contemporary Thinkers*, edited by Richard Kearney. New York: New York University Press, 1995.

Lila, Salvatore R.C. *Neuplatonisches Gedankengut.* In *"Homilien über die Seligpreisungen" Gregors von Nyssa.* Edited by Drobner. Leiden, 2004.
Louth, Andrew. *Maximus the Confessor*. London and New York: Routledge, 1996.
Ludlow, Morwenna. *Universal Salvation: Eschatology in the Thought of Gregory of Nyssa and Karl Rahner*. Oxford: Oxford University Press, 2000.
Manchester, Peter. *Temporality and Trinity*. New York: Fordham University Press, 2015.
Manoussakis, John Panteleimon. "Spelling Desire with Two Ls: Levinas and Lacan." In *JPCS: Journal for the Psychoanalysis of Culture & Society*, 7:1 (Spring 2002): 16–23.
Manoussakis, John Panteleimon. *God After Metaphysics: A Theological Aesthetics*. Bloomington: Indiana University Press, 2007.
Manoussakis, John Panteleimon and DeRoo, Neal (eds). *Phenomenology and Eschatology: Not Yet in the Now*. Surrey: Ashgate, 2009.
Manoussakis, John Panteleimon. "The Philosopher-Priest and the Mythology of Reason." In *Analecta Hermeneutica* 4 (2012): 1–18.
Manoussakis, John Panteleimon. "The Dialectic of Communion and Otherness in St. Maximus' Understanding of the Will." In *Knowing the Purpose of Creation Through the* Resurrection. Edited by Bishop Maxim (Vasiljević). Sebastian Press, 2013: 159–81.
Marion, Jean-Luc. *God Without Being*. Translated by Thomas A. Carlson. Chicago: The University of Chicago Press, 1991.
Marion, Jean-Luc. *Reduction and Givenness*. Translated by Thomas A. Carlson. Evanston: Northwestern University Press, 1998.
Marion, Jean-Luc. *Idol and the Distance*. Translated by Thomas A. Carlson. New York: Fordham University Press, 2001.
Marion, Jean-Luc. *Being Given*. Translated by Jeffrey L. Kosky. Stanford: Stanford University Press, 2002.
Marion, Jean-Luc. *In Excess: Studies of Saturated Phenomena*. Translated by Robyn Horner and Vincent Berraud. New York: Fordham University Press, 2002.
Marion, Jean-Luc. *Erotic Phenomenon*. Translated by Stephen E. Lewis. Chicago: The University of Chicago Press, 2007.
Marion, Jean-Luc. *In the Self's Place: The Approach of Saint Augustine*. Translated by Jeffrey L. Kosky. Stanford: Stanford University Press, 2012.
Martens, Paul. "Origen: Kierkegaard's Equivocal Appropriation of Origen of Alexandria." In *Kierkegaard and the Patristic and Medieval Traditions*, edited by Jon Stewart. Hampshire: Ashgate, 2008.
Martin, Dale. *The Corinthian Body*. New Haven: Yale University Press, 1995.
Maximus the Confessor. *Centuries on Theology and Economy*. In *Philocalia*, vol. II. Athens: Aster, 1984.
Maximus the Confessor. *On the Cosmic Mystery of Jesus Christ*. Translated by Paul M. Blowers and Robert Louis Wilken. Crestwood: St. Vladimir's Seminary Press, 2003.
Maximus the Confessor. *On Difficulties in the Church Fathers*. Translated by Nicholas Constas. Cambridge: Harvard University Press, 2014.
Merleau-Ponty, Maurice. *Le monde sensible et le monde de l'expression: Cours au Collège de France, Notes 1953*. Paris: Métis Presses, 2011.
Merleau-Ponty, Maurice. *Phenomenology of Perception*. Translated by Donald A. Landes. London: Routledge, 2012.
Mitralexis, Sotiris. *Ever-moving Repose: The Notion of Time in Maximus the Confessor's Philosophy Through the Perspective of Relational Ontology*. Berlin: Freie Universität Berlin, 2014.

Moyaert, Paul. "Lacan on Neighborly Love: The Relation to the Thing in the Other Who Is My Neighbor." In *Epoché* 4, 1 (1996): 1–31.
Nason, Shannon M. *Motion, Change, and Activity in the Thought of Søren Kierkegaard*. Unpublished thesis. Indiana: Purdue University, 2008.
Nietzsche, Friedrich. *Philosophy in the Tragic Age of the Greeks*. Translated by Marianne Cowan. Washington, DC: Regnery Publishing, 1962.
Nietzsche, Friedrich. *On the Genealogy of Morals*. Translated by Walter Kaufmann and R.J. Holingdale. New York: Vintage Books, 1967.
O'Donnell, James J. *Augustine Confessions*. Oxford: Oxford University Press, 2012.
Origen. *Commentariorum series in Matthaeum*. In *Origenes Werke* vol. 12. Edited by Erich Klostermann. Leipzig, 1942.
Origen. *De Principiis*. Translated by G.W. Butterworth. Gloucester: Peter Smith, 1973.
Papagiannopoulos, Ilias. *Ἐπέκεινα τῆς ἀπουσίας*, Athens: Indiktos, 2005.
Pieper, Joseph. *Divine Madness*. Translated by Lothar Krauth. San Francisco: Ignatius, 1995.
Pirandello, Luigi. *Undici Novelle*. Translated by Stanley Appelbaum. New York: Dover, 1994.
Plass, Paul, C. "'Moving Rest' in Maximus the Confessor." In *Classica et Mediaevalia*, 34 (1984): 177–90.
Plato: Complete Works. Edited by John M. Cooper. Indianapolis and Cambridge: Jackett, 1997.
Pound, Marcus. *Theology, Psychoanalysis and Trauma*. London: SCM Press, 2007.
Ratzinger, Joseph. *Eschatology: Death and the Eternal Life*. Translated by Michael Waldstein. Washington, DC: The Catholic University of America Press, 1988.
Richardson, William J. "Psychoanalysis and the God-Question." In *Thought: A Review of Culture and Idea*, 61: 240 (1986): 68–83.
Richardson, William, J. "'Like A Straw': Religion and Psychoanalysis." In *Eros and Eris*, edited by P.J.M. van Tongeren. The Hague: Kluwer Academic Publishers, 1992.
Ricoeur, Paul. *The Symbolism of Evil*. Translated by Emerson Buchanan. Boston: Beacon Press, 1967.
Ricoeur, Paul. *Freud & Philosophy: An Essay on Interpretation*. Translated by Denis Savage. New Haven and London: Yale University Press, 1970.
Ricoeur, Paul. "'Original Sin': A Study in Meaning" in *The Conflict of Interpretations*, edited by Don Ihde. Evanston: Northwestern University Press, 1974.
Ricoeur, Paul. *Time and Narrative*. Translated by Kathleen McLaughlin and David Pellauer. Chicago: The University of Chicago Press, 1984.
Sartre, Jean-Paul. *Being and Nothingness*. New York: Washington Square Press, 1956.
Sartre, Jean-Paul. *The Transcendence of the Ego*. Translated by Forrest Williams and Robert Kirkpatrick. New York: Hill and Wang, 1960.
Schelling, F.W.J. *Philosophical Investigations into the Essence of Human Freedom*. Translated by Jeff Love and Johannes Schmidt. New York: SUNY Press, 2006.
Scholten, Clemens. "Einleitung." In Philoponos, *De Aeternitate Mundi. Fontes Christiani*, 64/1. Turnhout: Brepols, 2009.
Severson, Eric. *Levinas's Philosophy of Time: Gift, Responsibility, Diachrony, Hope*. Pittsburg: Duquesne University Press, 2013.
Sherwood, Polycarp. *The Earlier Ambigua of Saint Maximus the Confessor and his Refutation of Origenism*. Rome: Pontificium Institutum S. Anselmi, 1955.
Sinclair, W.M. "Eutychius." In *Dictionary of Christian Biography and Literature to the End of Sixth Century*. Edited by William C. Piercy and Henry Wace. London: John Murray, 1880.

Sokolowski, Robert. *Introduction to Phenomenology*. Cambridge: Cambridge University Press, 2000.
Sophocles. *Oedipus Rex*. Translated by David Grene. In *Greek Tragedies*, vol. 1. Edited by D. Grene and R. Lattimore. Chicago: University of Chicago Press, 1942.
Sorabji, Richard. "The Concept of the Will from Plato to Maximus the Confessor." In *The Will and Human Action*, edited by Thomas Pink and M.W.F. Stone. London and New York: Routledge, 2004.
Thomas Aquinas. *Summa Theologiae*. Edited and translated by Timothy McDermott. Cambridge: Cambridge University Press, 2006.
Thuberg, Lars. *Microcosm and Mediator: The Theological Anthropology of Maximus the Confessor*. Chicago: Open Court, 1995.
Tzamalikos. *The Concept of Time in Origen*. Peter Lang, 1991.
Vessey, Mark. "Book Review, *Saint Augustine: Confessions*." In *Augustinian Studies* 24 (1994): 163–81.
Virno, Paolo. *Déjà Vu and the End of History*. Translation by David Broder. London and New York: Verso, 2015.
Vlastos, Gregory. "Reasons and Causes in the *Phaedo*." In *The Philosophical Review*, 78:3 (1969): 291–325.
von Balthasar, Hans Urs. *Cosmic Liturgy: The Universe According to Maximus the Confessor*. Translated by Brian E. Daley. San Francisco: Ignatius, 1988.
von Balthasar, Hans Urs. *Presence and Thought: An Essay on the Religious Philosophy of Gregory of Nyssa*. Translated by Mark Sebanc. San Francisco: Ignatius, 1995.
von Balthasar, Hans Urs. *Theo-Drama V: The Last Act*. Translated by Graham Harrison. San Francisco: Ignatius Press, 1998.
von Balthasar, Hans Urs. *Theo-Logic, II: Truth of God*. Translated by Adrian J. Walker. San Francisco: Ignatius Press, 2004.
Ware, James. "Paul's Understanding of the Resurrection in 1 Corinthians 15:36–54." In *Journal of Biblical Literature*, 133:4 (2014): 809–835.
Warhol, Andy. *The Philosophy of Andy Warhol*. Harcourt, 1975.
Waugh, Evelyn. *Brideshead Revisited: The Sacred and Profane Memories of Captain Charles Ryder*. Boston: Back Bay Books, 1991.
Waugh, Evelyn. *Brideshead Revisited*. Boston: Back Bay Books, 1999.
Wetzel, James. "The Trappings of Woe and Confession of Grief." In *A Reader's Companion to Augustine's Confessions*. Edited by Kim Paffenroth and Robert P. Kennedy. London: Westminster John Knox Press, 2003.
White, Caroline. *Gregory of Nazianzus: Autobiographical Poems*. Cambridge: Cambridge University Press, 1996.
Wyschogrod, Edith. *Emmanuel Levinas: The Problem of Ethical Metaphysics*. New York: Fordham University Press, 2000.
Zizioulas, John, D. *Communion and Otherness*. London and New York: T&T Clark, 2006.
Zizioulas, John, D. *The Eucharistic Communion and the World*. Edited by Luke Ben Tallon. New York: Continuum, 2011.
Zizioulas, John, D. "Εσχατολογία καὶ Ὕπαρξη." In *Σύναξη* 121 (2012): 43–72.

Index

www.ingramcontent.com/pod-product-compliance
Ingram Content Group UK Ltd.
Pitfield, Milton Keynes, MK11 3LW, UK
UKHW020141290926
15471UKWH00004B/627

9 781350 101814